INSTALLATION EFFECTS IN GEOTECHNICAL ENGINEERING

PROCEEDINGS OF THE INTERNATIONAL CONFERENCE ON INSTALLATION EFFECTS IN GEOTECHNICAL ENGINEERING, ROTTERDAM, THE NETHERLANDS, 24–27 MARCH 2013

Installation Effects in Geotechnical Engineering

Editors

Michael A. Hicks
Section of Geo-Engineering, Delft University of Technology, Delft, The Netherlands

Jelke Dijkstra
Section of Geo-Engineering, Delft University of Technology, Delft, The Netherlands

Marti Lloret-Cabot
Centre for Geotechnical and Materials Modelling, University of Newcastle, Newcastle, Australia & Section of Geo-Engineering, Delft University of Technology, Delft, The Netherlands

Minna Karstunen
Department of Civil and Environmental Engineering, Chalmers University of Technology, Gothenburg, Sweden & University of Strathclyde, Glasgow, UK

CRC Press
Taylor & Francis Group
Boca Raton   London   New York   Leiden

CRC Press is an imprint of the
Taylor & Francis Group, an **informa** business

A BALKEMA BOOK

CRC Press/Balkema is an imprint of the Taylor & Francis Group, an informa business

© 2013 Taylor & Francis Group, London, UK

Typeset by V Publishing Solutions Pvt Ltd., Chennai, India

Published by: CRC Press/Balkema
    P.O. Box 11320, 2301 EH Leiden, The Netherlands
    e-mail: Pub.NL@taylorandfrancis.com
    www.crcpress.com – www.taylorandfrancis.com

ISBN: 978-1-138-00041-4 (Hbk + CD-ROM)
ISBN: 978-0-203-74654-7 (eBook)

Table of contents

Soil-structure interaction

Installation Effects in Geotechnical Engineering – Hicks et al. (eds)
© *2013 Taylor & Francis Group, London, ISBN 978-1-138-00041-4*

Preface

The partners of the European project GEO-INSTALL extend a warm welcome to all participants of the International Conference on Installation Effects in Geotechnical Engineering (ICIEGE). This is the closing conference of GEO-INSTALL (FP7/2007-2013, PIAG-GA-2009-230638), an Industry-Academia Pathways and Partnerships project funded by the European Community from the 7th Framework Programme.

Infrastructure construction involves the installation of structural elements, such as piles and various ground improvement techniques for soils and rocks. The installation process itself can be quasi-static (for example jacked piles) or dynamic (vibratory methods, such as stone columns and driven piles), and generally involves very large deformations and changes in pore pressure. The fact that natural soils are complex geomaterials, exhibiting structure and rate-dependent behaviour, makes analysis of such problems yet more challenging. In particular, the influence of installation on key design parameters, such as mobilised strength at the soil-structure interface and soil stiffness, is difficult to quantify and, as yet, impossible to model. Numerical analyses using the standard Finite Element Method (FEM) are unable to produce accurate descriptions of large deformation problems due to excessive mesh distortions and novel techniques need to be developed.

The aim of the conference is to provide an international forum for presenting the latest developments in monitoring, analysing and managing installation effects in geotechnical engineering. Active discussion on important topics will be facilitated through invited keynote lectures, which set the scene for the main theme of the conference. In addition, the partners of GEO-INSTALL will present selected highlights of their joint research programme, which has been achieved through intense collaboration between industry and academia.

The peer review papers contained in these proceedings were accepted for presentation at ICIEGE, held in Rotterdam, The Netherlands 24–27 March 2013. They have been authored by academics, researchers and practitioners from many countries worldwide and cover numerous important aspects related to installation effects in geotechnical engineering, ranging from large deformation modelling to real field applications. The main topics are:

- Computational methods
- Constitutive modelling
- Installation effects
- Offshore construction and foundations
- Soil improvement
- Soil-structure interaction

The submitted abstracts were reviewed and the authors of those abstracts that fell within the scope of the conference were invited to submit full papers for peer review. The editors would like to thank the Scientific Committee who provided assistance in the review process. They would also like to thank the keynote speakers, authors, participants and members of the Organising Committee. The editors are grateful for the support of the European Community and the partner organisations of the GEO-INSTALL project: University of Strathclyde, United Kingdom (Project Coordinator); Delft University of Technology, The Netherlands; Deltares, The Netherlands; Keller Limited, United Kingdom; Norwegian Geotechnical Institute, Norway; Plaxis BV, The Netherlands; Stellenbosch University, South Africa; University of Stuttgart, Germany.

On behalf of the partners of GEO-INSTALL, we welcome you to The Netherlands and hope that you find the conference both enjoyable and inspiring.

Michael Hicks
Jelke Dijkstra
Marti Lloret-Cabot
Minna Karstunen
January 2013

Committees

ORGANISING COMMITTEE

Prof. Michael Hicks, *Delft University of Technology, The Netherlands (Chair)*
Dr. Jelke Dijkstra, *Delft University of Technology, The Netherlands (Co-Chair)*
Dr. Marti Lloret-Cabot, *University of Newcastle, Australia & Delft University of Technology, The Netherlands*
Ms. Bahar Akbarian, *Delft University of Technology, The Netherlands*
Mr. Marius Ottolini, *Delft University of Technology, The Netherlands*
Mr. Remon Romp, *Delft University of Technology, The Netherlands*

SCIENTIFIC COMMITTEE

Dr. Lars Andresen, *Norwegian Geotechnical Institute, Norway*
Dr. Patrick Becker, *University of Strathclyde, UK*
Prof. Alan Bell, *Keller, UK*
Prof. Thomas Benz, *Norwegian University of Science and Technology, Norway*
Dr. Ronald Brinkgreve, *Plaxis, The Netherlands*
Dr. Mike Brown, *Dundee University, UK*
Dr. Jorge Castro, *University of Cantabria, Spain*
Dr. Corne Coetzee, *Stellenbosch University, South Africa*
Dr. Jelke Dijkstra, *Delft University of Technology, The Netherlands*
Dr. Derek Egan, *Keller, UK*
Dr. Gustav Grimstad, *Oslo and Akershus University College of Applied Sciences, Norway*
Dr. Claire Heaney, *Plaxis, The Netherlands*
Prof. Michael Hicks, *Delft University of Technology, The Netherlands*
Prof. Minna Karstunen, *Chalmers University of Technology, Sweden & Strathclyde University, UK (GEO-INSTALL Coordinator)*
Dr. Marti Lloret-Cabot, *University of Newcastle, Australia & Delft University of Technology, The Netherlands*
Mr. Dirk Luger, *Deltares, The Netherlands*
Prof. Cesar Sagaseta, *University of Cantabria, Spain*
Dr. Nallathamby Sivasithamparam, *Plaxis, The Netherlands*
Prof. Pieter Vermeer, *Deltares, The Netherlands & University of Stuttgart, Germany*
Dr. Jimmy Wehr, *Keller, Germany*
Prof. Zdizslaw Wieckowski, *Technical University of Lodz, Poland*
Prof. Zhenyu Yin, *Shanghai Jiao Tong University, China*

Installation Effects in Geotechnical Engineering – Hicks et al. (eds)
© *2013 Taylor & Francis Group, London, ISBN 978-1-138-00041-4*

Keynote speakers

Dr. Lars Andresen, *Norwegian Geotechnical Institute, Norway*
Prof. Minna Karstunen, *Chalmers University of Technology, Sweden & University of Strathclyde, UK*
Mr. Alain Puech, *Fugro, France*
Prof. Pieter Vermeer, *Deltares, The Netherlands*
Dr. Jimmy Wehr, *Keller, Germany*
Prof. David White, *University of Western Australia, Australia*

Computational methods

Installation Effects in Geotechnical Engineering – Hicks et al. (eds)
© *2013 Taylor & Francis Group, London, ISBN 978-1-138-00041-4*

Large deformation analysis of cone penetration testing in undrained clay

L. Beuth
Deltares, Delft, The Netherlands

P.A. Vermeer
Deltares, Delft, The Netherlands
University of Stuttgart, Germany

ABSTRACT

Cone penetration testing is a widely-used in-situ test for soil profiling as well as estimating soil properties of strength and stiffness. In this paper, the relationship between the undrained shear strength of clay and the measured cone tip resistance is investigated through numerical analysis. Such analyses serve to refine and establish correlations between cone penetration testing measurements and soil properties; thus enabling more reliable predictions of soil properties. The presented analyses are performed by means of a Material Point Method that has been developed specifically for the analysis of quasi-static geotechnical problems involving large deformations of soil. Both, the load-type dependency of the shear strength of undrained clay as well as the influence of the anisotropic fabric of natural clay on the undrained shear strength are taken into account through a new material model, the Anisotropic Undrained Clay model. Results indicate that the deformation mechanism relevant for cone penetration in undrained normally-consolidated clay differs significantly from predictions based on the Tresca model, but resulting cone factors appear to be useful.

REFERENCES

Beuth, L. 2012. *Formulation and application of a quasi-staticmaterial point method.* Ph.D. thesis, University of Stuttgart, Holzgartenstr. 16, 70174 Stuttgart.

Beuth, L., T. Benz, P. Vermeer, C. Coetzee, P. Bonnier, & P. Van Den Berg 2007. Formulation and validation of a quasistatic Material Point Method. In *Proceedings of the 10th International Symposium on Numerical Methods in Geomechanics,* Volume 10, pp. 189–195. Taylor & Francis Group.

Beuth, L., Z. Wi eckowski, & P. Vermeer 2011. Solution of quasi-static large-strain problems by the material point method. *International Journal for Numerical and Analytical Methods in Geomechanics 35*(13), 1451–1465.

Coetzee, C., P. Vermeer, & A. Basson 2005. The modelling of anchors using the material point method. *International Journal for Numerical and Analytical Methods in Geomechanics 29*(9), 879–895.

Guilkey, J. & J. Weiss 2003. Implicit time integration for the material point method: Quantitative and algorithmic comparisons with the finite element method. *International Journal for Numerical Methods in Engineering 57*(9), 1323–1338.

Lu, Q., M. Randolph, Y. Hu, & I. Bugarski. 2004. A numerical study of cone penetration in clay. *Géotechnique 54*(4), 257–267.

Schofield, A. & P. Wroth. 1968. *Critical state soil mechanics.* McGraw–Hill New York.

Sulsky, D., Z. Chen, & H. Schreyer 1994. A particle method for history–dependent materials. *Computer Methods in Applied Mechanics and Engineering 118*(1–2), 179–196.

Van den Berg, P. 1994. *Analysis of soil penetration.* Ph.D. thesis, Technical University Delft, The Netherlands.

Vermeer, P., I. Jassim, & F. Hamad 2010. Need and performance of a new undrained clay model.

Vermeer, P., Y. Yuan, L. Beuth, & P. Bonnier 2009. Application of interface elements with the Material Point Method. In *Proceedings of the 18th International Conference on Computer Methods in Mechanics,* Volume 18, pp. 477–478. Polish Academy of Sciences.

Wheeler, S., A. Näätänen, M. Karstunen, & M. Lojander 2003. An anisotropic elastoplastic model for soft clays. *Canadian Geotechnical Journal 40*(2), 403–418.

Więckowski, Z., S. Youn, & J. Yeon 1999. A particle–in–cell solution to the silo discharging problem. *International Journal for Numerical Methods in Engineering 45*(9), 1203–1225.

Installation Effects in Geotechnical Engineering – Hicks et al. (eds)
© 2013 Taylor & Francis Group, London, ISBN 978-1-138-00041-4

Adaptive Mesh Refinement for strain-softening materials in geomechanics

C.E. Heaney & R.B.J. Brinkgreve
Delft University of Technology, Delft, The Netherlands
Plaxis BV, Delft, The Netherlands

P.G. Bonnier
Plaxis BV, Delft, The Netherlands

M.A. Hicks
Geo-Engineering Section, Faculty of CITG, Delft University of Technology, Delft, The Netherlands

ABSTRACT

This paper describes the implementation of Adaptive Mesh Refinement (AMR) within the geotechnical software package Plaxis 2D. The algorithm is recovery-based and aims to reduce the discretization error estimated as a measure of the incremental deviatoric strain. Once the global discretisation error exceeds a user-defined tolerance, certain elements in the mesh are marked for refinement. The refinement procedure is based on a combination of regular subdivision and longest-edge bisection. Mapping from the old mesh to the new mesh is accomplished by using the recovered solutions at the nodes and the shape functions. The AMR algorithm is demonstrated for a vertical cut problem for a softening Drucker-Prager material. Regularisation prevents the mesh-dependency which would otherwise be seen for such softening constitutive models.

REFERENCES

Bank, R., Sherman, A.H., & Weisser, A. 1983. Refinement algorithms and data structures for regular local mesh refinement. In *Scientific computing (IMACS Transactions)*, North Holland, pp. 3–17.

Boroomand, B. & Zienkiewicz, O.C. 1999. Recovery procedures in error estimation and adaptivity. *Comput. Methods Appl. Mech. Engrg* 176, 127–146.

Brinkgreve, R.B.J. 1994. *Geomaterial models and numerical analysis of softening.* Ph. D. thesis, Delft University of Technology.

Brinkgreve, R.B.J., Swolfs, W.M., & Engin, E. 2011. *Plaxis 2D 2011.* Plaxis BV.

Heaney, C.E., Bonnier, P.G., Brinkgreve, R.B.J., & Hicks, M.A. 2013. Adaptive mesh refinement with application to geomaterials. *In preparation for submission to Comput. Geotech.*

Hicks, M.A. 2000. Coupled computations for an elastic perfectly plastic soil using adaptive mesh refinement. *Int. J. Numer. Anal. Meth. Geomech. 24*, 453–476.

Hu, Y. & Randolph, M.F. 1998. *H*-adaptive FE analysis of elasto-plastic non-homogeneous soil with large deformation.*Comput. Geotech. 23*, 61–83.

Kardani, M., Nazem, M., Abbo, A.J., Sheng, D., & Sloan, S.W. 2012. Refined h-adaptive FE procedure for large deformation geotechnical problems. *Comput. Mech. 49*, 21–33.

Mar, A. & Hicks, M.A. 1996. A benchmark computational study of finite element error estimation. *Int. J. Nume. Meth. Engng 39*(23), 3969–3983.

Peri´c, D., Hochard, C., Dutko, M., & Owen, D.R.J. 1996. Transfer operators for evolving meshes in small strain elasto-plasticity. *Comput. Methods Appl. Mech. Engrg 137*, 331–344.

Rezania, M., Bonnier, P.G., Brinkgreve, R.B.J., & Karstunen, M. 2012. Non-local regularisation of Drucker-Prager softening model. In Z. Yang (Ed.), *Proceedings of the 20th UK National Conference of ACME*, Manchester (UK), pp. 275–278.

Rivara, M.C. 1984. Design and data structures of a fully adaptive multigrid finite element software. *ACM Transactionson Mathematical Software 10*, 242–264.

Rolshoven, S. & Jir´asek, M. 2003. Numerical aspects of nonlocal plasticity with strain softening. In *Computational Modelling of Concrete Structures*, Austria, pp. 305–314.

Rosenberg, I.G. & Stenger, F. 1975. A lower bound on the angles of triangles constructed by bisecting the longest side. *Mathematics of Computation 29*, 390–395.

Zienkiewicz, O.C. & Zhu, J.Z. 1987. A simple error estimator and adaptive procedure for practical engineering analysis. *Int. J. Numer. Meth. Engng 24*, 337–357.

Zienkiewicz, O.C. & Zhu, J.Z. 1992a. The superconvergent patch recovery and *a posteriori* error estimates. *Int. J. Numer. Meth. Engng 33*, 1331–1364.

Zienkiewicz, O.C. & Zhu, J.Z. 1992b. The superconvergent patch recovery (SPR) and adaptive finite element refinement. *Comput. Methods Appl. Mech. Engrg 101*, 207–224.

Zienkiewicz, O.C. & Zhu, J.Z. 1995. Superconvergence and the superconvergent patch recovery. *Finite Elem. Anal. Des. 19*, 11–23.

A dynamic material point method for geomechanics

I. Jassim
Institut für Geotechnik, Universität Stuttgart, Germany

C. Coetzee
Department of Mechanical Engineering, University of Stellenbosch, South Africa

P.A. Vermeer
Deltares, Delft, The Netherlands
University of Stuttgart, Germany

ABSTRACT

A dynamic Material Point Method for use in Geomechanics is presented. Soil and structural bodies are represented by (material) particles, which move inside an unstructured mesh of four-noded 3-D tetrahedral elements. As such low-order elements tend to show locking for fully developed plastic flow, a strain-enhancement remedy is described. As a first example, the penetration of a drop anchor into a Mohr-Coulomb soil is considered. As both a soil body and a metal anchor are considered, an algorithm for dynamic contact is used and described. An improved type of absorbing boundaries to avoid the reflection of stress waves is also described. The second example consists of dynamic cone penetration. Finally, the example of a collapsing tunnel is considered.

REFERENCES

Bardenhagen, S.G. Brackbill, J.U. & Sulsky, D. 2000, "Thematerial-point method for granular materials". Computer Methods in Applied Mechanics and Engineering, Vol. (187), 529–541.

Bathe, K.J. 1982, Finite Elements Procedures in Engineering Analysis, Prentice-Hall, Inc., New Jersey.

Belytschko, T. Lu, Y.Y. & Gu, L. 1994, "Element-free Galerkin methods". International Journal of Numerical Methods in Engineering, Vol. (37), 229–256.

Beuth, L., Więckowski, Z. & Vermeer, P. 2010, "Solution of quasi-static large-strain problems with the material point method". International Journal of Numerical and Analytical Methods in Geomechanics, Wiley Online Library (wileyonlinelibrary.com). DOI: 10.1002/nag.965.

Borja, R.I. 1988, "Dynamics of pile driving by the finite element method". Computers and Geotechnics, Vol. (5), 39–49.

Burgess, D. Sulsky, D. & Brackbill, J.U. 1992, "Massmatrix formulation of the FLIP particle-in-cell method". Journal of Computational Physics, Vol. (103), 1–15.

Coetzee, C.J. Vermeer, P.A. & Basson, A.H. 2005, "The modelling of anchors using the material point method". International Journal for Numerical and Analytical Methods in Geomechanics, Vol. (29), 879–895.

Detournay, C. & Dzik E. 2006, "Nodal mixed discretization for tetrahedral elements", Proceeding of '4 international FLAC symposium on numerical modeling in geomechanics, Itasca Consulting Group.

Harlow, F.H. 1964, "The particle-in-cell computing method for fluid dynamics". Methods for Computational Physics, Vol. (3), 319–343.

Lysmer, J. & Kuhlmeyer, R.L. 1969, "Finite dynamic model for infinite media". Journal of the Engineering Mechanics Division, Vol. (95), 859–877.

Sulsky, D., Zhou, S.J. & Schreyer, H.L. 1995, "Application of aparticle-in-cell method to solid mechanics". Computer Physics Communications, Vol. (87), 236–252.

Sulsky, D. & Schreyer, H.L. 1996, "Axisymmetric form of the material point method with applications to upsetting and Taylor impact problems". Computer Methods in Applied Mechanics and Engineering, Vol. (139), 409–429.

Więckowski, Z., Youn, S.K. & Yeon, J.H. 1999, "A particle-in-cell solution to the silo discharging problem". International Journal for Numerical Methods in Engineering, Vol. (45), 1203–1225.

Więckowski, Z. 2004, "The material point method in large strain engineering problems". Computer Methods in Applied Mechanics and Engineering, Vol. (193), 4417–4438.

Pile penetration simulation with Material Point Method

L.J. Lim & A. Andreykiv
Plaxis BV, Delft, The Netherlands

R.B.J. Brinkgreve
Delft University of Technology, Delft, The Netherlands
Plaxis BV, Delft, The Netherlands

ABSTRACT

Conventional Finite Element Method (FEM) faces mesh distortion and mesh tangling when it is used in the simulation of extreme deformation in pile penetration. To avoid the shortcoming of FEM, Material Point Method (MPM) is used owing its ability to nalyse engineering problems involving extreme deformation. However, MPM generates numerical noise in the calculation of stresses when material points cross element boundary due to discontinuity of gradient of shape functions. Dual Domain Material Point Method (DDMP), introduced earlier within an explicit framework, provides a continuity of the gradient of shape functions which helps to eliminate the numerical noise in stress and strain fields. In this paper we further extended the application of DDMP within an implicit scheme by formulating a consistent tangent system. Additionally, we have presented a method to couple MPM and FEM analyses in order to limit the application of MPM to the areas with extreme deformation, which allows to increase computational efficiency. Numerical analysis results for a pile penetration problem have been presented and compared with analytical solution for validation.

REFERENCES

Bardenhagen, S.G. & Kober, E.M. 2004. The generalied interpolation material point method. *Computer Modeling in Engineering and Sciences* 5(3), 477–495.

Das, B.M. 2007. *Principles of Foundation Engineering.* Cengage Publisher. Geuzaine, C. & Remacle, J.F. 2009. Gmsh: a three dimensional finite element mesh generator with built-in pre- and post-processing facilities. *International Journal for Numerical Methods in Engineering 79*(79), 1309–1331.

Harlow, F.H. 1964. The particle-in-cell computing method for fluid dynamics. *Methods in Computational Physics 3*, 319.

Sadeghirad, A., Brannon, R.M., & Burghardt, J. 2011. A convected particle domain interpolation technique to extend applicability of the material point method for problems involving massive deformations. *International Journal for Numerical Methods in Engineering 86*, 1435–1456.

Sulsky, D., Zhou, S., & Schreyer, H. 1995. Application of a particle-in-cell method to solid mechanics. *Computer Physics Communications 87*, 236.

Wells, G.N. 2009. *The Finite Element Method: An Introduction.* University of Cambridge and Delft University of Technology.

Wieckowski, Z. 2004. The material point method in large strain engineering problems. *Computer Methods in Applied Mechanics and Engineering 193*, 4417–4438.

Zhang, D.Z., Ma, X., & Giguere, P.T. 2011. Material point method enhanced by modified gradient of shape function. *Journal of Computational Physics 230*, 379–6398.

Installation Effects in Geotechnical Engineering – Hicks et al. (eds)
© 2013 Taylor & Francis Group, London, ISBN 978-1-138-00041-4

Coupling triangular plate and volume elements in analysis of geotechnical problems

S. Tan
Geo-Engineering Section, Faculty of CITG, Delft University of Technology, Delft, The Netherlands

M.A. Hicks
Geo-Engineering Section, Faculty of CITG, Delft University of Technology, Delft, The Netherlands
Deltares, Delft, The Netherlands

A. Rohe
Deltares, Delft, The Netherlands

ABSTRACT

In geotechnics, it is common to have a thin layered material with high stiffness on top of soil, to prevent damage from external loading or erosion. To model this numerically a very fine mesh is often needed, which decreases the critical time step in explicit time integration algorithms, severely affecting simulation performance. The use of 2D plate elements connected to 3D elements is investigated to overcome this problem. The presented three-noded plate element is based on Kirchhoff thin plate theory, and uses non-conforming polynomial shape functions. The lumped mass matrix, with both the translational and rotational degrees of freedom considered, is implemented with an explicit time integration scheme. This plate element is coupled with volume elements and the implementation is tested for several cases in which analytical or numerical solutions are available. All simulations show that the plate element with the lumped mass matrix is working properly in geotechnical problems.

REFERENCES

Batoz, J.L., Bathe, K.J. & Ho, L.W. 1980. A study of three-node triangular plate bending elements. *International Journal for Numerical Methods in Engineering*, 15: 1771–1812.

Bauchau, O.A. & Craig, J.I. 2009. *Structural analysis with applications to aerospace structures*. Springer, The Netherlands.

Bazeley, G.P., Cheung, Y.K., Irons, B.M. & Zienkiewicz, O.C. 1965. Triangular elements in plate bending conforming and non-conforming solutions. *Proceedings Conference on Matrix Methods in Structural Mechanics*, Wright Patterson A.F.B., Ohio. 547–576.

Burmister, D.M. 1945. The general theory of stresses and displacements in layered systems. *International Journal of Applied Physics*, 16: 89–94.

Burmister, D.M. 1958. Evaluation of pavement systems of the WASHO road test by layered systems methods. *Highway Research Board Bulletin 177*.

FLAC 1998. Fast lagrangian analysis of continua: theory and background. Itasca Consultin Group, Inc., Minnesota, USA.

Reddy, J.N. 2007. Theory and analysis of elastic plates and shells. CRC, Taylor and Francis.

Specht, B. 1988. Modified shape functions for the three-node plate bending element passing the patch test. *International Journal for Numerical Methods in Engineering*, 26: 705–715.

Surana, K.S. 1978. Lumped mass matrices with non-zero inertia for general shell and axisymmetric shell elements. *International Journal for Numerical Methods in Engineering*, 12: 1635–1650.

Yoder, E.J. & Witczak, M.W. 1975. *Principles of pavement design* (2nd Edition). John Wiley & Sons.

Installation Effects in Geotechnical Engineering – Hicks et al. (eds)
© 2013 Taylor & Francis Group, London, ISBN 978-1-138-00041-4

Fracture growth in heterogeneous geomaterials

P.J. Vardon & J.D. Nuttall

Department of Geoscience and Engineering, Delft University of Technology, Delft, The Netherlands

ABSTRACT

A method to simulate fracture growth in hetero-geneous geomaterials is presented. The method links a statistical description of spatial material properties, random fields, with the extended finite element method. A probabilistic description of the growing fracture is then generated which can be incorporated into risk and reliability design based methods. Initial model development, verification and output is presented.

REFERENCES

Belytschko, T. & Black, T. 1999. Elastic crack growth in finite elements with minimal remeshing. *Int. J., Numer. Meth. Engng.* 45: 601–620.

Bordas, S., Nguyen, P.V., Dunant, C., Guidoum, A. & Nguyen-Dang, H. 2007.An extended finite element library. *Int. J. Num. Meth. Engng.* 71(6): 703–732.

Fenton, G.A. & Vanmarcke, E.H. 1990. Simulation of random fields via Local Average Subdivision. *ASCE J. Eng. Mech.* 116(8): 1733–49.

Hicks, M.A., editor. 2007. *Risk and variability in geotechnical engineering.* London: Thomas Telford.

Hicks, M.A. & Samy, K.. 2002. Influence of heterogeneity on undrained clay slope stability. *Quart. J. Eng. Geol. Hydrogeol.* 35(1): 41–9.

Hicks, M.A. & Spencer, W.A. 2010. Influence of heterogeneity on the reliability and failure of a long 3D slope. *Computers and Geotechnics* 37(7–8): 948–955.

Moes, N., Dolbow, J. & Belytschko, T. 1999. A finite element method for crack growth without remeshing. *Int. J. Numer. Meth. Engng.* 46: 131–150.

Pais, M.J. 2010. *MATLAB eXtended Finite Element Method (MXFEM): User guide.* Gainesville, Florida: University of Florida.

Shih, C. & Asaro, R. 1988. Elastic-plastic analysis of cracks on biomaterial interfaces: part I—small scale yielding. *J. App. Mech.* 55: 299–316.

Constitutive modelling

Modelling rate-dependent behaviour of structured clays

M. Karstunen
Chalmers University of Technology, Gothenburg, Sweden
University of Strathclyde, Glasgow, Scotland, UK

N. Sivasithamparam
Plaxis BV, Delft, The Netherlands
University of Strathclyde, Glasgow, Scotland, UK

R.B.J. Brinkgreve
Plaxis BV, Delft, The Netherlands
Delft University of Technology, Delft, The Netherlands

P.G. Bonnier
Plaxis BV, Delft, The Netherlands

ABSTRACT

Due to the desire of reducing the embedded CO_2 in construction and the pressure in public finances to get more value for money in big infrastructure projects, the demands for the accuracy of deformation predictions increase. Instead of piling, alternative environmentally friendly and cost effective solutions, such as preloading via surcharge, vertical drains and column methods, such as deep-mixing, are becoming increasingly attractive. Installation of piles and ground improvement into the ground will modify the state of the soil. This is sometimes beneficial, and sometimes detrimental, and so far this effect has been rarely taken into account. One reason for this is that the numerical techniques and the constitutive soil models have not been able to represent the changes in soil structure and state in a satisfactory manner. The aim of the European GEO-INSTALL project has been to develop numerical techniques that can be used to model installation effects in geotechnical engineering. A key part of this has been constitutive model development, and their robust implementation. The aim of this paper is to discuss some recently developed rate-dependent constitutive models for structured clays, which formed the basis for new developments, resulting in a new rate-dependent model able to represent the complex rate-dependent stress-strain behaviour of soft structured clays. The importance of modelling key features of soil behaviour in the context of rate-dependency are discussed in the light of experimental evidence, and demonstrated through a series of numerical benchmark simulations.

REFERENCES

Bai X. & Smart P. 1996. Change in microstructure of kaolin in consolidation and undrained shear. Géotechnique 47(5): 1009–1017.

Bodas Freitas, T.M., Potts, D.M. & Zdravkovic L. 2011. A time-dependent constitutive model for soils with isotach viscosity. Computers and Geotechnics 38(6): 809–820.

Brinkgreve, R.B.J., Swolfs, W.M. & Engin, E. 2010. Plaxis 2D 2010 Manual.

Burland, J.B. 1990. On the compressibility and shear strength of natural clays. Géotechnique 40(3): 329–378.

Castro, J. & Karstunen, M. 2010. Numerical simulations of stone column installation. Canadian Geotechnical Journal 47:1127–1138.

Castro, J., Karstunen, M., Sivasithamparam, N. & Sagaseta C. 2013. Numerical analyses of stone column installation in Bothkennar clay. Proc. International Conference on Installation Effects in Geotechnical Engineering, Rotterdam, NL, 23–27 March 2013.

Dijkstra, J., Alderliests, E.A. & Broere, W. 2011. Photoelastic investigation into plugging of open ended piles. Frontiers in Offshore Geotechnics II: 483–488. CRC Press. DOI: 10.1201/b10132-61.

Dijkstra, J. & Broere, W. 2010. New full-field stress measurement method using photoelasticity. Geotechnical Testing Journal 33(3). DOI: 10.1520/GTJ102672.

Gens, A. & Nova, R. 1993. Conceptual bases for a constitutive model for bonded soils and weak rocks. In Geotechnical Engineering of Hard Soils—Soft Rocks, Athens, Greece, Anagnostopoulos et al. (eds.). Balkema, Rotterdam: 485–494.

Graham, J., Crooks., J.H.A. & Bell, A.L. 1983. Time effects on the stress strain behaviour of natural soft clays. Géotechnique 33(3): 327–340.

Grimstad, G., Abate, S., Nordal, S. & Karstunen, M. 2010. Modelling creep and rate effects in structured anisotropic soft clays. Acta Geotechnica 5: 69–81.

Hicher, P.Y., Wahyudi, H. & Tessier, D. 2000. Microstructural analysis of inherent and induced anisotropy in clay. Mechanics of Cohesive-Frictional Materials 5(5): 341–371.

Hinchberger, S.D. & Qu, G. 2009. Viscoplastic constitutive approach for rate-sensitive structured clays. Canadian Geotech. J. 46(6): 609–626.

Karstunen, M. & Koskinen, M. 2004. Anisotropy and destructuration of Murro clay. Proc., Advances in Geotechn. Eng. Skempton Conf., London, UK, 1, 476–487.

Karstunen, M. & Koskinen M. 2008. Plastic anisotropy of soft reconstituted clays. Canadian Geotechnical Journal 45: 314–328.

Karstunen, M., Krenn, H., Wheeler, S.J. Koskinen, M. & Zentar, R. 2005. Effect of anisotropy and destructuration on the behaviour of Murro test embankment. ASCE International Journal of Geomechanics 5(2): 87–97.

Karstunen, M., Rezania, M, Sivasithamparam, S. & Yin, Z.-Y. (in press). Comparison of anisotropic rate-dependent model for modelling consolidation of soft clays. International Journal of Geomechanics doi:10.1061/(ASCE)GM.1943-5622.0000267.

Karstunen, M. & Yin, Z.-Y. 2010. Modelling time-dependent behaviour of Murro test embankment. Géotechnique 60(10): 735–749.

Leoni, M., Karstunen, M. & Vermeer, P.A. 2008. Anisotropic creep model for soft soils. Géotechnique, 58(3): 215–226.

Leroueil S., Tavenas F., Brucy F., La Rochelle P. & Roy M. 1979. Behaviour of destructured natural clays. ASCE Journal of Geotechnical Engineering, 105(6): 759–778.

Leroueil. S. & Vaughan, P.R. 1990. The general and congruent effects of structure in natural soils and weak rocks. Géotechnique, 40(3):467–488.

Lobo-Guerrero, S. & Vallejo, L.E. 2005. DEM analysis of crushing around driven piles in granular materials. Géotechnique 55(8): 617–623.

Perzyna, P. 1963. The constitutive equations for work-hardening and rate sensitive plastic materials. Proc. Vibration Problems Warsaw 3, 281–290.

Pusch, R. 1970. Clay microstructure. A study of the microstructure of soft clays with special reference to their physical properties. Swedish Geotechnical Institute, Proceedings No. 24. Stockholm.

Pusch R. Personal communication, 2012.

Rankka, K. 2003. Kviklera-billdning och egenskaper. SGI Varia 526. Swedish Geotechnical Institute, Linköping.

Rankka, K., Andersson-Sköld, Y. Hulten C., Larsson, R. Lerocex, V. & Dahlin, T. 2004. Quick clays in Sweden SGI Report 65. Swedish Geotechnical Institute, Linköping.

SGI. 1995. Geotekniska skadekostnader och behov av ökad geoteknisk kunskap. Internal Report K94/1825/3. In Swedish, not publicly available.

Sheng, D., Sloan, S. & Yu, H. 2000. Aspects of finite element implementation of critical state models. Computational Mechanics 26: 185–196.

Sivasithamparam, N. 2012. Modelling creep behaviour of soft soils. Internal report. Plaxis B.V. & University of Strathclyde, not publicly available.

Sivasithamparam, N., Karstunen, M., Brinkgreve, R.B.J.& Bonnier P.G. 2013. Comparison of two anisotropic rate dependent models at element level. Proc. International Conference on Installation Effects in Geotechnical Engineering, Rotterdam, NL, 23–27 March 2013.

Symposium. 1992. Bothkennar soft clay test site: Characterization and lessons learned (Géotechnique symposium in print). Géotechnique 42(2): 161–380.

Vermeer P.A. & Neher H.P. 1999. A soft soil model that accounts for creep. Proc. Int. Symp. Beyond 2000 in Computational Geotechnics. Amsterdam. Balkema, Rotterdam, 249–261.

Vermeer, P.A., Stolle D.F.E. & Bonnier P.G. 1998. From the classical theory of secondary compression to modern creep analysis. Proc. Computer Methods and Advances in Geomechanics. Balkema, Rotterdam.

Wheeler, S.J., Näätänen A., Karstunen M. & Lojander M. 2003. An anisotropic elasto-plastic model for natural soft clays. Canadian Geotechnical Journal 40(2): 403–418.

Yin, Z.-Y., Karstunen, M., Chang, C.S., Koskinen, M., & Lojander, M. 2011. Modelling time-dependent behaviour of soft sensitive clay. J. Geotech. Geoenviron. Eng., 137(11): 1103–1113.

Installation Effects in Geotechnical Engineering – Hicks et al. (eds)
© 2013 Taylor & Francis Group, London, ISBN 978-1-138-00041-4

FEM simulation of large vertical deformations caused by land subsidence and verification of the results by using radar interferometry techniques

C. Loupasakis & D. Rozos
Laboratory of Engineering Geology and Hydrogeology, School of Mining and Metallurgical Engineering, National Technical University of Athens, Athens, Greece

F. Raspini & S. Moretti
Department of Earth Sciences, University of Firenze, Firenze, Italy

ABSTRACT

Land subsidence induced by the over-exploitation of aquifers is a very common phenomenon affecting extensive areas worldwide. Knowledge about the range and the rate of the deformations is necessary for the installation and the protection of constructions. A detailed study was conducted aiming to validate the efficiency of the Mohr–Coulomb and the Hardening soil models, introduced in the PLAXIS 2D finite-element code, for the simulation of land subsidence. The data used for this case study came from the Kalochori region on the west side of Thessaloniki, Northern Greece. The rapid development of the area and the extensive need of water led to the development of surface subsidence, reaching maximum values of 3–4 m, in several parts of the study area. The validation of the simulation results was conducted by using all available subsidence indications as well as a land motion mapping produced by PSI (Persistent Scatterer Interferometry) analysis.

REFERENCES

Andronopoulos, V. 1979. Geological and geotechnical study in the Kalochori (Thessaloniki) area. Institute of Geology and Mineral Exploration Report, Athens, p. 90.

Andronopoulos, V., Rozos, D. & Hatzinakos I. 1990. Geotechnical study of ground settlement in the Kalochori area, Thessaloniki District. Institute of Geology and Mineral Exploration Report, Athens, p. 45.

Andronopoulos, V., Rozos, D. & Hatzinakos, I. 1991. Subsidence phenomena in the industrial area of Thessaloniki, Greece. In: Johnson, A. (Ed.). Land Subsidence, pp. 59–69.

Brinkgreve, R.B.J., Al-Khoury, R. & Bakker, K.J. 2002. Plaxis, Fine Element Code for Soil and Rock Analysis, 2D–Version 8. Balkema, Rotterdam.

Colesanti, C., Ferretti, A., Prati, C. & Rocca, F., 2003. Monitoring landslides and tectonic motions with the Permanent Scatterers Technique. Engineering Geology (68) pp. 3–14.

Ferretti, A., Prati, C. & Rocca, F., 2000. Nonlinear subsidence rate estimation using Permanent Scatterers in differential SAR interferometry. IEEE Trans. Geosci. Remote Sens. (38) pp. 2202–2212.

Ferretti, A., Prati, C. & Rocca, F., 2001. Permanent Scatterers in SAR interferometry. IEEE Trans. Geosci. Remote Sens. (39) pp. 8–20.

Hatzinakos, I., Rozos, D. & Apostolidis, E. 1990. Engineering geological mapping and related geotechnical problems in the wider industrial area of Thessaloniki, Greece. In: Price, D. (Ed.), Proceedings of Sixth International IAEG Congress, Amsterdam, Balkema, pp. 127–134.

Loupasakis, C. & Rozos, D., 2009. Land Subsidence Induced by Water Pumping in Kalochori Village (North Greece)—Simulation of the Phenomenon by Means of the Finite Element Method, Quarterly Journal of Engineering Geology and Hydrogeology, Geological Society of London, (42) pp. 369–382.

Massonnet, D., Rossi, M., Carmona, C., Adragna, F., Peltzer, G., Feigl, K. & Rabaute, T., 1993. The displacement field of the Landers earthquake mapped by radar interferometry. Nature (364).

Massonnet, D., Feigl, K.L., Rossi, M. & Adragna, F. 1994. Radar interferometric mapping of deformation in the year after the Landers earthquake. Nature (369).

Massonnet, D. & Feigl, K.L., 1995. Discrimination of geophysical phenomena in satellite radar interferograms. Geophys. Res. Lett., (22).

Massonnet, D., Briole, P. & Arnaud, A. 1995. Deflation of Mount Etna monitored by Spaceborne Radar Interferometry. Nature, (375).

Massonnet, D., Thatcher, W. & Vadon, H. 1996. Detection of postseismic fault zone collapse following the Landers earthquake. Nature (382).

Massonnet, D. & Feigl, K.L., 1998. Radar interferometry and its application to changes in the Earth's surface. Rev. Geophys., (36).

Mouratidis A., Briole P., Astaras A., Pavlidis S., Tsakiri M., Ilieva M., Rolandone F. & Katsambalos, K. 2010. Contribution of InSAR and kinematic Gps data to subsidence and geohazard monitoring in Central Macedonia (N. Greece), Scientific Annals, School of Geology, Aristotle University of Thessaloniki, Proceedings of the XIX CBGA Congress, Thessaloniki, Greece (100) pp. 535–545.

Nicolau, S. & Nicolaidis, M. 1987. Geoelectric study in Kalochori village of Thessaloniki. Report, Institute of Geology and Mineral Exploration, Athens, 10.

Psimoulis P., Ghilardi M., Fouache E. & Stiros S., 2007. Subsidence and evolution of the Thessaloniki plain, Greece, based on historical leveling and GPS data. Engineering Geology (90) pp. 55–70.

Rozos, D. & Hadzinakos, I. 1993. Geological conditions and geomechanical behaviour of the neogene sediments in the area west of Thessaloniki (Greece). Proc. Int. Symp. on Geotechnical Engineering of Hard Soils—Soft Rocks, Greece, Anagnostopoulos et al (Eds), A.A. Balkema, Vol.1, pp. 269–274, 1993 Rotterdam.

Rozos, D. Apostolidis, E. & Hadzinakos, I. 2004. Engineering geological map of the wider Thessaloniki area, Greece. Bulletin of Intern. Assoc. of Eng. Geol. & the Environment, Springer-Verlag, Vol. 63, pp. 103–108, 2004. Rotterdam.

Rosen, P.A., Hensley, S., Joughin, I.R., Li, F.K., Madsen, S.N., Rodriguez, E. & Goldstein, R.M., 2000. Synthetic aperture radar interferometry. Proc. I.E.E.E.(88).

Soulios, G. 1999. Research for the development of the aquifers in the low lands on the west of Thessaloniki for the interests of the Water Company of Thessaloniki. unpublished technical report, Aristotle University of Thessaloniki, p. 99.

Stiros, S.C. 2001. Subsidence of the Thessaloniki (northern Greece) coastal plain, 1960–1999. Engineering Geology, (61) pp. 243–256.

Singhroy, V., Mattar, K.E. & Gray, A.L. 1998. Landslide characterisation in Canada using interferometric SAR and combined SAR and TM images. Advances in Space Research (21) pp. 465–476.

Tralli, D.M, Blom, R.G., Zlotnicki, V., Donnellan, A., Evans, D.L. 2005. Satellite remote sensing of earthquake, volcano, flood, landslide and coastal inundation hazards. Journal of Photogrammetry and Remote Sensing (59) pp. 185–198.

Zebker, H.A. & Goldstein, R.M. 1986. Topographic Mapping From Interferometric Synthetic Aperture Radar Observations. Journal of Geophysical Research (91), pp. 4993–4999.

Zebker, H.A. & Villasenor, J. 1992. Decorrelation in interferometric radar echoes. IEEE Trans. Geosci. Remote Sens. (30).

Installation Effects in Geotechnical Engineering – Hicks et al. (eds)
© *2013 Taylor & Francis Group, London, ISBN 978-1-138-00041-4*

Numerical modeling of fracturing in soil mix material

G. Van Lysebetten & A. Vervoort
KU Leuven, Leuven, Belgium

N. Denies & N. Huybrechts
Belgian Building Research Institute, Geotechnical division, Limelette, Belgium

J. Maertens
Jan Maertens BVBA
KU Leuven, Leuven, Belgium

F. De Cock
Geotechnical Expert Office GEO.BE, Lennik, Belgium

B. Lameire
Belgian Association of Foundation Contractors ABEF, Brussels, Belgium

ABSTRACT

The deep soil mixing technique consists of an in situ mechanical mixing of the soil with an injected binder (e.g. cement). However, the presence of soil inclusions (poorly or even unmixed soil) in the artificial material is unavoidable. This heterogeneous character of soil mix material makes it different from traditional building materials. The presented research investigates the influence of the volume percentage of inclusions on the strength, stiffness, stress-strain behaviour and fracture pattern of soil mix material. 2D numerical simulations are conducted using a discrete element program (UDEC) and the results are compared with experimental data. It is observed that the reduction of the strength and stiffness of a sample is significantly larger than the weighted average of the UCS and Young's modulus, taking into account the volumes of the well mixed material and the softer inclusions. However, the strength is remarkably more affected by the volume percentage of inclusions than the stiffness. Moreover, other parameters than the percentage of weak material are also important (e.g. shape, size and relative position of the inclusions) and result in wide ranges of resulting strength and stiffness.

REFERENCES

Barber C.B., Dobkin D.P., Huhdanpaa, H. 1996. The Quickhull algorithm for convex hulls. *ACM Transactions on Mathematical Software*, Vol. 22 (4), pp. 469–483.

Cundall P.A. 1971. A computer model for simulating progressive large scale movements in block rock systems. *Proceedings of the International Society for Rock Mechanics Symposium*, paper II–8.

Cundall P.A. & Board M. 1988. A microcomputer program for modeling large-strain plasticity problems. *Proceedings of the 6th International Conference on Numerical Methods in Geomechanics. Rotterdam (Netherlands)*, pp. 2101–2108.

Debecker B. 2009. Influence of planar heterogeneities on the fracture behavior of rock, *Ph.D. Dissertation, KU Leuven, Leuven (Belgium)*.

Debecker B., Vervoort A., Napier J.A.L. 2006. Fracturing in and around a natural discontinuity in rock: a comparison between boundary and discrete element models. *Proceedings of the 5th International Conference on Engineering Computational Technology, Las Palmas de Gran Canaria (Spain)*, paper 168.

Debecker B. & Vervoort A. 2006. A 2D triangular Delaunay grid generator for the simulation of rock features. *Proceedings of the 5th International Conference on Engineering Computational Technology, Las Palmas de Gran Canaria (Spain)*, paper 220.

Denies N., Huybrechts N., De Cock F., Lameire B., Maertens J., Vervoort A. 2012a. Soil mix walls as retaining structures, Belgian practice. *Proceedings of TC 211 International Symposium on Ground Improvement, Brussels (Belgium)*, Vol. 3, pp. 83–97.

Denies N., Huybrechts N., De Cock F., Lameire B., Vervoort A., Maertens J. 2012b. Soil mix walls as retaining structures, mechanical characterization. *Proceedings of the TC 211 International Symposium on Ground Improvement, Brussels (Belgium)*, Vol. 3, pp. 99–115.

Ganne P., Denies N., Huybrechts N., Vervoort A., Tavallali A., Maertens J., Lameire B., De Cock F. 2011. Soil Mix: influence of soil inclusions on the structural behavior. *Proceedings of the 15th European conference on soil mechanics and geotechnical engineering, Athens (Greece)*, pp. 977–982.

Itasca 2004. UDEC v4.0 manual. Itasca Consulting Group, Inc., Minnesota, USA.

Potyondy D.O. & Cundall P.A. 2004. A Bonded-Particle Model for Rock. *International Journal for Rock Mechanics & Mineral Sciences*, Vol. 41(8), pp. 1329–1364.

Tempone P. & Lavrov A. 2008. DEM modeling of mudlosses into single fractures and fracture networks. *Proceedings of the 12th international conference of the international association for computer methods and advances in geomechanics, Goa (India)*, pp. 2475–2482.

Van Lysebetten G., Vervoort A., Maertens J., Huybrechts N. 2012. Discrete modelling for the study of the effect of soft inclusions on the behaviour of soil mix material. *In preparation.*

Van Lysebetten G. 2011. Soil Mix for Construction purposes: Quality control. *M.Sc. Thesis (Geotechnical and Mining Engineering) KU Leuven, Leuven.*

Vervoort A., Van Lysebetten G., Tavallali A. 2012. Numerical modeling of fracturing around soft inclusions. *Proceedings of the Southern Hemisphere International Rock Mechanics Symposium, Sun City (South-Africa)*, pp. 33–46.

A 3D practical constitutive model for predicting seismic liquefaction in sands

A. Petalas & V. Galavi
Plaxis BV, Delft, The Netherlands

R.B.J. Brinkgreve
Plaxis BV, Delft, The Netherlands
Delft University of Technology, Delft, The Netherlands

ABSTRACT

This paper presents a three dimensional formulation of a simple and practical constitutive model developed for evaluating seismic liquefaction in sands. The model is an extension of the two dimensional UBCSAND model developed at University of British Colombia (Beaty & Byrne 1998) which utilises isotropic and simplified kinematic hardening rules for primary and secondary yield surfaces, in order to take into account the effect of soil densification and predict a smooth transition into the liquefied state during undrained cyclic loading. By means of a simplified Rowe stress-dilatancy theory the model is capable of modelling cyclic liquefaction for different stress paths. To show the capability of the model to predict cyclic liquefaction in soils,the mechanical behaviour of some sands is numerically studied under direct simple shear conditions and compared with experimental data. The effect of densification and parameters selection on the results is discussed.

Finally, the model is used for simulating an experimental dynamic centrifuge test and the numerical results are compared with the real measurements.

REFERENCES

Beaty, M. & Byrne, P. 1998. An effective stress model for predicting liquefaction behaviour of sand. *Geotechnical Earthquake Engineering and Soil Dynamics III ASCE Geotechnical Special Publication No.75. 1*, 766–777.

Beaty, M. & Byrne, P. 2011. Ubcsand constitutive model version 904ar. *Itasca UDM Web Site*, 69.

Byrne, P.M., Park, S.S., Beaty, M., Sharp, M., Gonzales, L. & Abdoun, T. 2004. Numerical modelling of dynamic centrifuge tests. *13th World Conference in Earthquake Engineering.*

Martin, G., Finn, W., & Seed, H. 1975. Fundamentals of liquefaction under cyclic loading. *Journal of the Geotechnical Engineering Division, ASCE 101.*

Petalas, A., Galavi, V., & Bringkreve, R. 2012. Validation and verification of a practical constitutive model for predicting liquefaction in sands. *Proceedings of the 22nd European Young Geotechnical Engineers Conference, Gothenburg, Sweden.*, 167–172.

Puebla, H., Byrne, M., & Phillips, P. 1997. Analysis of canlex liquefaction embankments prototype and centrifuge models. *Canadian Geotechnical Journal 34*, 641–657.

Rowe, P.W. 1962. The stress-dilatancy relation for static equilibrium of an assembly of particles in contact. *Proceedings of the Royal Society of London. Series A, Mathematical and Physical Sciences 269A*, 500–527.

Sriskandakumar, S. 2004. Cyclic loading response of frasersand for validation of numerical models simulating centrifuge tests. *Master's thesis, The University of British Columbia, Department of Civil Engineering.*

Tsegaye, A. 2010. Plaxis liquefaction model. Report No. 1. *PLAXIS knowledge base.*

Installation Effects in Geotechnical Engineering – Hicks et al. (eds)
© 2013 Taylor & Francis Group, London, ISBN 978-1-138-00041-4

Comparison of two anisotropic creep models at element level

N. Sivasithamparam
Plaxis BV, Delft, The Netherlands
University of Strathclyde, Glasgow, Scotland, UK

M. Karstunen
Chalmers University of Technology, Gothenburg, Sweden
University of Strathclyde, Glasgow, Scotland, UK

R.B.J. Brinkgreve
Plaxis BV, Delft, The Netherlands
Delft University of Technology, Delft, The Netherlands

P.G. Bonnier
Plaxis BV, Delft, The Netherlands

ABSTRACT

This paper presents a comparison of two Anisotropic Creep Models, ACM and Creep-SCLAY1, which differ in their formulation of creep strain rate. Creep is formulated in ACM using the concept of contours of constant volumetric creep strain rate, whereas the newly developed Creep-SCLAY1 model uses the concept of a constant rate of visco-plastic multiplier. The two models are identical in the way the initial anisotropy and the evolution of anisotropy are simulated. A key assumption of both models is that there is no purely elasticdomain. The models are compared at element level. The numerical simulations show that the Creep-SCLAY1 model is able to give a better representation of natural clay behaviour at element level.

REFERENCES

Brinkgreve, R., Engin, E., & Swolfs, W.M. 2012. *PLAXIS Finite Element Code for Soil and Rock Analyses*. The Netherlands: 2D-Version 2011.

Graham, J., Crooks, J., & Bell, A. 1983. Time effects on the stress-strain behaviour of natural soft clays. *Geotechnique 33(3)*, 327–340.

Grimstad, G. 2009. *Development of effective stress based anisotropic models for soft clays*. Ph.D. thesis, Norwegian University of Science and Technology (NTNU), Norway.

Grimstad, G., Abate, S., Nordal, S., & Karstunen, M. 2010. Modeling creep and rate effects in structured anisotropic soft clays. *Acta Geotechnica 5*, 69–81.

Karstunen, M., Krenn, H., Wheeler, S., Koskinen, M., & Zentar, R. 2005. The effect of anisotropy and destructuration on the behaviour of Murro test embankment. *Int. J. of Geomechanics (ASCE) 5(2)*, 87–97.

Karstunen, M. & Yin, Z.Y. 2010. Modelling timedependent behaviour of Murro test embankment. *Geotechnique 29*, 1–34.

Leoni, M., Karstunen, M., & Vermeer, P. 2008. Anisotropic creep model for soft soils. *Gotechnique 58(3)*, 215–226.

Leroueil, S. & Marques, M. 1996. State of art: Importance of strain rate and temperature effects in geotechnical engineering. Measuring and modelling time dependent behaviour of soils. *ASCE, Geotechnical Special Publication 61*, 1–60.

Roscoe, K. & Burland, J. 1968. On the generalised stress-strain behaviour of wet clay. *Engineering Plasticirv*, 535–609.

Sekiguchi, H. & Ohta, H. 1977. Induced anisotropy and time dependency in clays. *9th ICSMFE, Tokyo, Constitutive equations of Soils 17*, 229–238.

Sivasithamparam, N. 2012. Modelling creep behaviour of soft soils. *Internal report Plaxis B.V.* Symposium 1992. Bothkennar soft clay test site: Characterization and lessons learned (Géotechnique symposium in print). *Geotechnique 42(2)*, 161–380.

Tatsuoka, F., Ishihara, M., Di Benedetto, H., & Kuwano, R. 2002. Time-dependent shear deformation characteristics of geomaterials and their simulation. *Soils & Foundations 42(2)*, 103–138.

Tavenas, F., Leroueil, S., La Rochelle, P., & Roy, M. 1978. Creep behaviour of an undisturbed lightly overconsolidated clay. *Can. Geot. J. 15(3)*, 402–423.

Vaid, Y. & Campanella, R. 1977. Time-dependent behaviour of undisturbed clay. *ASCE J Geotech Eng Div 103(7)*, 693–709.

Vermeer, P.A. & Neher, H. 1999. A soft soil model that accounts for creep. *Beyond 2000 in Computational Geotechnics, R.B.J. Brinkgreve(eds), Rotterdam. 4*, 249–261.

Vermeer, P.A., Stolle, D.F.E., & Bonnier, P.G. 1998. From classical theory of secondary compression to modern creep analysis. *Proc. 9th Int. Conf. Comp. Meth. and Adv. Geomech., Yuan(eds) 4*, 2469–2478.

Wheeler, S., Näätänen, A., Karstunen, M., & Lojander, M. 2003. An anisotropic elasto-plastic model for soft clays. *Can. Geot. J. 40*, 403–418.

Yin, Z.Y., Chang, C.S., Karstunen, M., & Hicher, P.Y. 2010. An anisotropic elastic-viscoplastic model for soft clays. *Int. J. of Solids and Structures 47*, 665–677.

Zentar, R., Karstunen, M., Wiltafafsky, C., Schweiger, H.F., & Koskinen, M. 2002. Comparison of two approaches for modelling anisotropy of soft clays. *Proc. 8th Int. Symp. on Numerical Models in Geomech. NUMOG VIII*, 115–121.

Zhou, C., Yin, J.-H., Zhu, J.-G., & Cheng, C.-M. 2006. Elastic anisotropic viscoplastic modeling of the strain-rate dependent stress–strain behaviour of K0-consolidated natural marine clays in triaxial shear test. *Int. J. Geomech 5(3)*, 218–232.

Installation effects

Displacement pile installation effects in sand

A. Beijer Lundberg, J. Dijkstra & A.F. van Tol
Delft University of Technology, Delft, The Netherlands

ABSTRACT

Installation of jacked or driven displacement piles imposes large deformations in the soil. These installation effects will influence the subsequent load-deformation response of the installed piles and should be taken into account in the description of the soil-structure interaction. A series of model pile tests were carried out in the geotechnical centrifuge at TU Delft. The tests combined horizontal contact stress measurements on an instrumented model pile and visual observation of the soil deformation adjacent to the pile. The model scaling, experimental set-up and results of pile installation in samples with various initial relative densities are discussed. The results indicated good consistency and reproduced the friction fatigue effect, the lateral loads at rest acting on the pile show surprisingly small influences from the pile installation.

REFERENCES

Allersma, H.G.B. 1994. The university of Delft Geotechnical centrifuge, *Proc. Int. Conf 1994.*

Axelsson, Gary. 1998. Long-term increase in shaft capacity of driven piles in sand, *Proc. 4th Int. Conf. on Case Histories in Geotech. Engng., St. Louis, Missouri.*

Beijer Lundberg, A., Djikstra, J. & van Beek, K. 2012. Measurements of soil contact stress in a harsh environment, *Instrumentation and Measurement Technology Conference (I2MTC), 2012 IEEE International. IEEE, 2012.*

Boulon, M. & Foray, P. 1986. Physical and numerical simulation of lateral shaft friction along offshore piles in sand, *In numerical methods in offshore pileing, 3rd international conference, Nantes, 1986.*

DeJong, J.T., White, D.J. & Randolph, M.F. 2006. Microscale observation and modeling of soil-structure interface behavior using particle image velocimetry, *Soils and Foundations,* 46(1): 15–28.

De Nicola, A. & Randolph, M.F. 1993. Tensile and compressive shaft capacity of piles in sand, *Journal of geotechnical engineering* 119(12), 1952–1973.

Eslami, A. & Fellenius, B.H. 1997. Pile capacity by direct CPT and CPTu methods applied to 102 case histories. *Canadian Geotechnical Journal* 34(6): 886–904.

Houlsby, G.T. 1991. How the dilatancy of soils affects their behavior, *Proceedings of the 10th European Conference on Soil Mechanics and Foundation Engineering.*

Huy, N.Q., Dijkstra, J. & Tol, A.F. van, 2005. Influence of loading rate on the bearing capacity of piles in sand. *Proceedings of the 16th International Conference on Soil Mechanics and Geotechnical Engineering,* pp. 2125–2128.

Jardine, R.J., Zhu, B.T., Foray, P. & Yang, Z.X. 2012. Measurements of stresses around closed ended displacement piles in sand, *Geotechnique,* 56(9), 1–17.

Klotz, E.U. & Coop, M.R. 2001. An investigation of the effect of soil state on the capacity of driven piles in sands, *Geotechnique,* 51(9), 733–751.

Kraft, L.M. 1991. Performance of axially loaded pipe piles in sand, *J. of Geotech. Engrg,* 117(2), 272–296.

Lee, J.H. & R. Salgado, 1999. Determination of pile base resistance in sands, *Journal of Geotechnical and Geoenvironmental Engineering* 125(8), 673–683.

Lehane, B.M., Jardine, R., Bond, A. & Frank, R. 1993. Mechanisms of Shaft friction in Sand from instrumented pile tests, *J. of Geotech. Engrg.* 119(1), 1–19.

Lehane, B.M., Schneider, J.A. & Xu, X. 2005. The UWA-05 method for prediction of axial capacity of driven piles in sand, *Proc., 1st Int. Symp. on Frontiers in Offshore Geotechnics. Perth, Australia: Balkema.*

Randolph, M.F., Dolwyn, J. & Beck, R. 1994. Design of driven piles in sand, *Geotechnique,* 44(3), 427–448.

Vennemann, P., Lindken, R. & Westerweel, J. 2007. In vivo whole-field blood velocity measurement techniques, *Experiments in fluids* 42(4), 495–511.

White, D.J. 2005. A general framework for shaft resistance on displacement piles in sand, *Proceedings of the 1st International Symposium on Frontiers in Offshore Geotechnics, ISFOG 2005. Taylor and Francis.*

White, D.J. & Lehane, B.M. 2004. Friction fatigue on displacement piles in sand, *Géotechnique* 54(10), 645–658.

Installation Effects in Geotechnical Engineering – Hicks et al. (eds)
© 2013 Taylor & Francis Group, London, ISBN 978-1-138-00041-4

Cyclic jacking of piles in silt and sand

F. Burali d'Arezzo & S.K. Haigh
University of Cambridge, Cambridge, UK

Y. Ishihara
Giken Seisakusho Co. Ltd, Kochi, Japan

ABSTRACT

Jacked piles are becoming a valuable installation method due to the low noise and vibration involved in the installation procedure. Cyclic jacking may be used in an attempt to decrease the required installation force. Small scale models of jacked piles were tested in sand and silt in a 10 m beam centrifuge. Two different piles were tested: smooth and rough. Piles were driven in two ways with monotonic and cyclically jacked installations. The cyclically jacked installation involves displacement reversal at certain depth for a fixed number of cycles. The depth of reversal and amplitude of the cycle vary for different tests. Data show that the base resistance increases during cyclic jacking due to soil compaction at the pile toe. On the other hand, shaft load decreases with the number of cycles applied due to densification of soil next to the pile shaft. Cyclic jacking may be used in unplugged tubular piles to decrease the required installation load.

REFERENCES

Balachowski, L. 2006. Scale effect in shaft friction from the direct shear interface tests. Archives of Civil and *Mechanical Engineering* VI(3).

Fioravante, V. 2002. On the shaft friction modelling of nondisplacement piles in sand. *Soils and foundations* 42(2), 23–33.

Gui, M. & M. Bolton, 1998. Guidelines for cone penetration tests in sand. *Centrifuge 98, Kimura, Kusakaba and Takemura*, 155–160.

Haigh, S., N. Houghton, & S. Lam 2010. Development of a 2D servo-actuator for novel centrifuge modelling. (2001), 239–244.

Mortara, G. 2007. Cyclic shear stress degradation and postcyclic behaviour from sandsteel interface direct shear tests. *Canadian Geotechnical Journal.* 44, 739–752.

Schofield, A.N. 1980, January. Cambridge Geotechnical Centrifuge Operations. *Géotechnique 30*(3), 227–268.

Silva, M. 2005. Numerical and physical models of rate effectsin soil penetration, Cambridge.

Taylor, R. 1995. Geotechnical Centrifuge Technology. *Blackie Academic and Professional.*

Uesugi, M. & H. Kishida 1987, January. Tests of the interface between sand and steel in the simple shear apparatus. *Géotechnique 37*(1), 45–52.

White, D. 2003. PSD measurement using the single particleoptical sizing (SPOS) method. *Géotechnique* 53, 317–326.

White, D. 2005. A general framework for shaft resistance ondisplacement piles in sand. *Proceedings of the InternationalSymposium, on Frontiers in Offshore Geotechnic*, 19–21 Sept 2005, Perth, WA, Australia.

White, D.J. & M.D. Bolton 2004, January. Displacement and strain paths during plane-strain model pile installation in sand. *Géotechnique 54*(6), 375–397.

Influence of installation procedures on the response of capacitance water content sensors

M. Caruso & F. Avanzi
Politecnico di Milano, Milano, Italy

C. Jommi
Politecnico di Milano, Milano, Italy
Delft University of Technology, Delft, The Netherlands

ABSTRACT

The effects of installation procedures on the performance of capacitance probes for monitoring water content changes in surficial soils are analysed numerically. The sensors measure the resonant frequency of an inductive-capacitive circuit, which includes the surrounding soil as capacitive element. Literature calibration curves are most often adopted to determine the soil water content from the measured resonant frequency. Alternatively, specific calibration for a given soil is performed in the laboratory. In both cases, reference is made to ideal conditions, in which perfect contact is assured between the probes access tube and the soil sample, prepared at uniform void ratio. Installation procedures in the field affect the system and they may hinder correct estimation of volumetric water content in the field. Numerical results are presented to quantify the influence of soil density changes promoted by installation operations and of the saturated kaolin/cement paste used to guarantee continuous contact between the access tube and the soil.

REFERENCES

Charlesworth, P. 2005. Soil water monitoring. An information package. 2nd ed. *Irrigation Insight* No. 1.

Dijkstra, J., Broere, W. & Van Tol, A.F. 2012. Electrical resistivity method for the measurement of density changes near a probe. *Géotechnique* 62(8): 721–725.

Gardner, C.M.K., Bell, J.P., Cooper, J.D., Dean, T.J., Hodnett, M.G. & Gardner, N. 1991. Soil Water Content. In Smith R.A., Mullings C.E., (eds) *Soil analysis—Physical methods*. Marcel Dekker, New York.

Hilhorst, M.A., Dirksen, C., Kampers, F.W.H., Feddes, R.A. 2000. New dielectric mixture equation for porous materials based on depolarization factors. *Soil Science Society of America Journal*. 64: 1581–1587.

Kelleners, T.J., Soppe, R.W.O., Robinson, D.A., Schaap, M.G., Ayars, J.E. & Skaggs, T.H. 2004. Calibration of capacitance probe sensors using electric circuit theory. *Soil Science Society of America Journal*. 68: 430–439.

Mojid, M.A., Wyseure, G.C.L. & Rose, D.A. 2003. Electrical conductivity problems associated with time-domain reflectometry (TDR) measurement in geotechnical engineering. *Geotechnical and Geological Engineering* 21: 243–258.

Nichol, C., Smith, L. & Beckie, R. 2003. Long-term measurement of matric suction using thermal conductivity sensors. *Canadian Geotechnical Journal*, 40: 587–597.

Noborio, K. 2001. Measurement of soil water content and electrical conductivity by time domain reflectometry: a review. *Comp and Elec in Agric* 31: 213–237.

Nörtemann, K., Hilland, J. & Kaatze, U. 1997. Dielectric properties of aqueous NaCl solutions at microwave frequencies. *J. Phys. Chem.* 101: 6864–6869.

Paltineanu, I.C. & Starr, J.L. 1997. Real time soil water dynamics using multisensors capacitance probes. *Soil Science Society of America Journal* 61: 1576–1585.

Randoplh, M.F., Carter, J.P. & Wroth, C.P. 1979. Driven piles in clay—the effects of installation and subsequent consolidation. *Géotechnique* 29(4): 361–393.

Robinson, D.A., Gardner, C.M.K., Evans, J., Cooper, J.D., Hodnett, M.G. & Bell, J.P. 1998. The dielectric calibration of capacitance probes for soil hydrology using an oscillation frequency response model. *Hydrology and Earth System Sciences* 2(1): 111–120.

Scanlon, B.R, Andraski, B.J. & Bilskie, J. 2002. Miscellaneous methods for measuring matric or water potential. In Dane, J.H. & Topp, G.C. (eds), *Methods of soil analysis, part 4, physical methods*. Soil Sci Soc. Am: 643–670.

Schwank, M. & Green, T.R. 2007. Simulated effects of Soil Temperature and Salinity on Capacitance Sensor Measurements. *Sensors*, 7(4): 548–577.

Schwank, M., Green, T.R., Mätzler, C., Benedickter, H. & Flürer, H. 2006. Laboratory characterization of a commercial capacitance sensor for estimating permittivity and inferring soil water content. *Vadose Zone Journal*. 5: 1048–1064.

Sentek 2001. Calibration of Sentek Pty Ltd Soil Moisture Sensors. Sentek Pty Ltd, Stepney, South Australia.

Topp, G.C., Davis, J.L. & Annan, A.P. 1980. Electromagnetic determination in soil-water content: measurement in coaxial transmission lines. *Water Resources Research*. 16: 574–582.

Installation Effects in Geotechnical Engineering – Hicks et al. (eds)
© *2013 Taylor & Francis Group, London, ISBN 978-1-138-00041-4*

The load capacity of driven cast in-situ piles derived from installation parameters

D. Egan
Keller Foundations, UK

ABSTRACT

The theoretical estimation of the ultimate capacity and service ability performance of common displacement foundation systems such driven cast in-situ (DCIS) piles is difficult due to the huge disruption in soil structure and in-situ stress regime caused by the installation process. Even though much research effort is expended on complex numerical modeling and reduced scale laboratory or centrifuge modeling there remains the difficulty of translating the knowledge gained into practical prediction tools appropriate for routine design and installation of the full size product in the field. So to advance and validate the conclusions drawn from numerical and small scale research the third strand of measuring and analysing full size field behavior must be added. This paper will summarise recent advances made in the field measurement and analysis of installation parameters to predict the load capacity of driven cast in-situ piles. The results from installation and testing of a DCIS pile is used to illustrate the methodology now being routinely by Keller Foundations in the UK. The conclusions drawn from this paper are already raising the standard of reliability, efficiency and sustainability of DCIS piles on routine projects.

REFERENCES

EN 12699:2001 Execution of special geotechnical works—Displacement Piles. 2001. BSI. London.

Evers, G., Hass, G., Frossard, A., Bustamante, M., Borel, S. & Skinner, H. 2003. Comparative performances of continuous flight auger and driven cast in place piles in sands. Deep foundations on bored and auger piles. Van Impe (ed) Millpress, Rotterdam.

Flynn, K., McCabe, B.A. & Egan, D. 2012. *Proceedings 9th International Conference on testing and design methods for deep foundations, Kanazawa, Japan.*

ICE Specification for Piling and Embedded Retaining Walls 2007. 2nd Ed. ICE. London.

Lunne T., Robertson P.K. & Powell, J.J.M 1997 Cone Penetration testing in geotechnical Practice.

Neely, W.J. 1990 bearing capacity of expanded-base piles with compacted concrete shafts. Journal of geotechnical Engineering, 116(9): 1309–1324.

Oasys Pile 19.2 User manual 2012 http://www.oasys-software.com/media/Manuals/Latest_Manuals/ Pile19.2_manual.pdf.

Installation Effects in Geotechnical Engineering – Hicks et al. (eds)
© 2013 Taylor & Francis Group, London, ISBN 978-1-138-00041-4

On the numerical modelling and incorporation of installation effects of jacked piles: A practical approach

H.K. Engin
Geo-Engineering Section, Delft University of Technology, Delft, The Netherlands

R.B.J. Brinkgreve
Geo-Engineering Section, Delft University of Technology, Delft, The Netherlands
Plaxis BV, Delft, The Netherlands

A.F. van Tol
Geo-Engineering Section, Delft University of Technology, Delft, The Netherlands
Deltares, Delft, The Netherlands

ABSTRACT

The installation process of a displacement pile causes a considerable amount of soil displacement and high levels of stresses, and therefore alters the soil state and properties around the pile. These installation effects may have important consequences on the performance of the pile in its service life (e.g. load—displacement behaviour) and on the neighbourhood (e.g. vibrations, nuisance). A more realistic behaviour and therefore an improved design would be achieved by considering the installation effects in the analyses. In current practice, the installation effects are taken into account by some empirical design methods in order to estimate the bearing capacity of foundation piles. The objective of this numerical study is to investigate and model the installation effects of pile jacking in sand in a numerical framework. In the first part of the study a simplified numerical technique was employed to investigate the installation effects. The results are approximated by nonlinear regression. Despite the limitations and simplifications, it was shown that the installation effects can be represented in terms of functional forms reasonably well. Furthermore, these functions can be easily applied in a standard FE analysis.

REFERENCES

Baligh, M.M. 1975. Theory of deep static cone penetration-resistance. Technical report, MIT Dept. of Civil Eng.

Bauer, E. 1996. Calibration of a comprehensive hypoplastic model for granular materials. *Japanese Geotechnical Society* 36(1), 13–26.

Broere, W. & A.F. van Tol 2006. Modelling the bearing capacity of displacement piles in sand. *In Proceedings of the ICE—Geotechnical Engineering*, Volume 159, pp. 195–206.

Chow, F.C. 1996. Investigations into the behaviour of displacement piles for offshore foundations. Ph.D. thesis.

Dijkstra, J. 2009. On the Modelling of Pile Installation. Ph.D.thesis.

Einav, I. & M.F. Randolph 2005. Combining upper boundand strain path methods for evaluating penetration resistance. *Int. J. Num. Meth. Engng.* 63(14), 1991–2016.

Engin, H.K., R.B.J. Brinkgreve, & A.F. van Tol 2011. Numerical analysis of installation effects of pile jacking insand. In G. Pietruszczak, S. & Pande (Ed.), *In proceedings of International Symposium on Computational Geomechanics, ComGeo-II*, pp. 744–755.

Lehane, B. 1992. Experimental investigations of pile behaviourusing instrumented field piles. Ph.D. thesis.

Liu, W. 2010. Axisymmetric centrifuge modelling of deep penetrationin sand. Ph.D. thesis.

Mahutka, K.P., F. König, & J. Grabe 2006. Numerical modelling of pile jacking, driving and vibratory driving. *In Proceedings of International Conference on Numerical Simulations of Construction Processes in Geotechnical Engineering for Urban Environment (NSC06)*, pp. 235–246.

Pham, H.D., H.K. Engin, R.B.J. Brinkgreve, & A.F. van Tol 2010. Modelling of installation effects of driven piles using hypoplasticity. In T. Benz and S. Nordal (Eds.), *Numerical Method in Geotechnical Engineering*, Numerical Methods in Geotechnical Engineering 2010: Proceedings of the Seventh European Conference on Numerical Methods in Geotechnical Engineering, pp. 261–266. Taylor & Francis.

Said, I., V. De Gennaro, & R. Frank 2008. Axisymmetric finiteelement analysis of pile loading tests. *Computers and Geotechnics* 36(1–2), 6–19.

Teh, C.I. & G.T. Houlsby 1991. An analytical study of the cone penetration test in clay. *Géotechnique* 41(1), 17–34.

van Langen, H. 1991. *Numerical Analysis of Soil-Structure Interaction* Ph.D. thesis.

vonWolffersdorff, P.A. 1996. A hypoplastic relation for granular materials with a predefined limit state surface. Mechanics of Cohesive-frictional Materials 1(3), 251–271.

White, D.J. 2002. *An investigation into the behaviour of pressed-in piles.* Ph.D. thesis

Analytical and laboratory study of soil disturbance caused by mandrel driven prefabricated vertical drains

A. Ghandeharioon

Department of Civil Engineering, Faculty of Science and Engineering, Laval University,
Quebec City, Quebec, Canada

ABSTRACT

Analytical investigations and large-scale laboratory experiments were conducted to study the soil disturbance due to the installation of mandrel-driven Prefabricated Vertical Drains (PVDs) in soft saturated clays. Considering the mandrels commonly used for installing PVDs in the field, an elliptical Cavity Expansion Theory (CET) was formulated to analyze the shear strain and pore pressure developed in soil during the mandrel installation. The elliptical CET was developed using modified Cam clay parameters for the undrained analysis of PVDs installed in soft soil deposits. This formulation identifies a critical zone, a plastic zone, and an elastic zone around mandrel-driven prefabricated vertical drains. The large-scale laboratory tests consider the effects of in-situ stresses using a specially designed consolidometer, and a rate controlled installation machine. The pore water pressure was measured at various locations during the installation of a PVD and withdrawal of the mandrel. The analytically predicted pore pressures agreed with the measurements in the laboratory. The results of moisture content tests were also analyzed to verify the concept of an elliptical smear zone around drains, and to derive a relationship between in-situ effective stresses and the extent of the smear zone.

REFERENCES

Bergado, D.T., Asakami, H., Alfaro, M.C. & Balasubramaniam, A.S. 1991. Smear effects of vertical drains on soft Bangkok clay, *Journal of Geotechnical Engineering, ASCE,* 117(10): 1509–1530.

Cao, L.F., Teh, C.I. & Chang, M.F. (2001). Undrained cavity expansion in modified Cam clay I: Theoretical analysis, *Geotechnique,* 51(4): 232–334.

Ghandeharioon, A. 2010. Analytical and numerical study of soil disturbance associated with the installation of mandrel driven prefabricated vertical drains, Ph.D. thesis, School of Civil, Mining & Environmental Engineering, Faculty of Engineering, Univ. of Wollongong, Wollongong, Australia.

Ghandeharioon, A. 2012. Large-scale laboratory assessment of smear effects in soft soils stabilized by prefabricated vertical drains. *Proceedings of 65th Canadian Geotechnical Conference: GeoManitoba—Building on the past,* Canadian Geotechnical Society, Canada.

Ghandeharioon, A., Indraratna, B. & Rujikiatkamjorn, C. 2010. Analysis of soil disturbance associated with mandrel driven prefabricated vertical drains using an elliptical cavity expansion theory, *International Journal of Geomechanics,* 10(2): 53–64.

Ghandeharioon, A., Indraratna, B. & Rujikiatkamjorn, C. 2012. Laboratory and finite element investigation of soil disturbance associated with the installation of mandrel driven prefabricated vertical drains, *Journal of Geotechnical and Geoenvironmental Engineering,* 138(3): 295–308.

Gibson, R.E. & Anderson, W.F. 1961. In-situ measurement of soil properties with the pressuremeter. *Civil Engineering Public Works Reviews,* 56: 615–618.

Hird, C.C. & Moseley, V.J. 2000. Model study of seepage in smear zones around vertical drains in layered soil, *Géotechnique,* 50(1): 89–97.

Indraratna, B. & Redana, I.W. 1998. Laboratory determination of smear zone due to vertical drain installation. *Journal of Geotechnical and Geoenvironmental Engineering,* 124(2): 180–184.

Sathananthan, I. & Indraratna, B. 2006. Laboratory evaluation of smear zone and correlation between permeability and moisture content, *Journal of Geotechnical and Geoenvironmental Engineering,* 132(7): 942–945.

Sharma, J.S. & Xiao, D. 2000. Characterisation of a smear zone around vertical drains by large-scale laboratory tests, *Canadian Geotechnical Journal,* 37: 1265–1271.

Vesic, A.S. 1972. Expansion of cavities in infinite soil mass. *Journal of the Soil Mechanics and Foundations Division, ASCE,* 98: 265–290.

Yu, H.S. & Houlsby, G.T. 1991. Finite cavity expansion in dilatant soil: Loading analysis. *Géotechnique,* 41, 173–183.

Yu, H.S. & Mitchell, J.K. 1996. Analysis of cone resistance: A review of methods. *The University of Newcastle, Australia,* Report No. 142.09.1996.

Installation Effects in Geotechnical Engineering – Hicks et al. (eds)
© *2013 Taylor & Francis Group, London, ISBN 978-1-138-00041-4*

CEL: Simulations for soil plugging, screwed pile installation and deep vibration compaction

J. Grabe, S. Henke, T. Pucker & T. Hamann
Institute of Geotechnical and Construction Engineering, Hamburg University of Technology, Hamburg, Germany

ABSTRACT

In many geotechnical applications like pile installation processes or soil improvement large deformations of the surrounding soil occur. These large deformations in combination with the complex material behaviour of the soil lead to numerical boundary value problems which are often difficult to solve. These difficulties are related to large mesh distortions and numerical problems due to complex contact conditions. One possibility to overcome these difficulties is to use special numerical techniques which are especially invented for large deformation simulations. In this work, the Coupled Eulerian-Lagrangian method (CEL) is identified to be well suited for such boundary value problems involving large distortions of the surrounding soil. Therefore, the numerical technique is presented and its suitability is shown in different geotechnical applications. The screwed pile installation and the deep vibration compaction process are investigated using the CEL method.

REFERENCES

Bienen B., Henke S. & Pucker T. 2011. Numerical study of the bearing behaviour of circular footings penetrating into sand. Proc. of 13th International Conference of International Association for Computer Methods and Advances in Geomechanics (IACMAG): 939–944.

Busch, P., Grabe, J., Gerressen, F.W. & Ulrich, G. 2010. Use of displacement piles for reinforcement of existing piles, Proceedings of DFI and EFFC 11th Int. Conf. in the DFI series, Geotechnical Challenges in Urban Regeneration in London/UK: 113–119.

Dassault Systèmes 2010. Abaqus User Manual Version 6.10.

Fellin W. 2000. Rütteldruckverdichtung als plastodynamisches Problem. Dissertation, Institute of Geotechnics and Tunneling, University of Innsbruck. Advances in Geotechnical Engineering and Tunneling, Heft 2, 2000.

Henke, S. 2008. Herstellungseinflüsse aus Pfahlrammung im Kaimauerbau, Dissertation, Veröffentlichungen des Instituts für Geotechnik und Baubetrieb der TU Hamburg-Harburg, Heft 18.

Henke, S., & Grabe, J. 2009. Numerical modeling of pile installation, Proc. of 17th Int. Conf. on Soil Mechanics and Foundation Engineering (ICSMFE): 1321–1324.

Henke S., Hamann T. & Grabe J. 2012. Numerische Untersuchungen zur Bodenverdichtung mittels Rütteldruckverfahren. 2. Symposium "Baugrundverbesserung in der Geotechnik", TU Wien, zur Veröffentlichung akzeptiert.

Henke S., Hamann T. & Grabe J. 2011.Coupled Eulerian-Lagrangian Simulation of the Deep Vibration Compaction Process as a Plastodynamic Problem. Proc. of EURODYN 2011, Leuven.

Henke, S., Qiu G. & Grabe J. 2010. A coupled eulerian-lagrangian approach to solve geotechnical problems involving large deformations. Proc. of 7th European Conference on Numerical Methods in Geotechnical Engineering (NUMGE) in Trondheim/Norway: 233–238.

Herle, I. 1997. Hypoplastizität und Granulometrie einfacher Korngerüste. Heft 142 Institut für Bodenmechanik und Felsmechanik der Universität Fridericana in Karlsruhe.

Mahutka, K.-P. 2007. Zur Verdichtung von rolligen Böden infolge dynamischer Pfahleinbringung und durch Oberflächenrüttler, Dissertation, Veröffentlichungen des Instituts für Geotechnik und Baubetrieb der TU Hamburg-Harburg, Heft 15.

Pichler, T., Pucker T., Hamann T., Henke S. & Qiu G. 2012. High-performance abaqus simulations in soil mechanics reloaded—chances and frontiers. Proc. of International Simulia Community Conference in Providense, Rhode Island/USA: 237–266.

Pucker, T. & Grabe J. 2012. Numerical simulation of the installation process of full displacement piles. Computers and Geotechnics, 45: 93–106, DOI: 10.1016/j.compgeo.2012.05.006.

Qiu, G. & Grabe J. 2011. Explicit modeling of cone and strip footing penetration under drained and undrained conditions using a visco-hypoplastic model. Geotechnik 34(3): 205–271.

Qiu, G., Henke S. & Grabe J. 2009. Applications of coupled eulerian lagrangian method to geotechnical problems with large deformations. Proc of SIMULIA Customer Conference 2009 in London: 420–435.

Qiu, G., S. Henke, & J. Grabe 2010. Application of a coupled eulerian-lagrangian approch on geomechanical problems involving large deformation. Computers and Geotechnics, DOI:10.1016/j.compgeo.2010.09.002.

vonWolffersdorff, P.-A. 1996. A hypoplastic relation for granular materials with a predefined limit state surface. Mechanics of Frictional and Cohesive Materials, 1: 251–271.

Witt K.J. 2009. Grundbau-Taschenbuch, Teil 2: Geotechnische Verfahren. Berlin, Ernst & Sohn Verlag.

Installation Effects in Geotechnical Engineering – Hicks et al. (eds)
© 2013 Taylor & Francis Group, London, ISBN 978-1-138-00041-4

Towards a framework for the prediction of installation rate effects

S. Robinson & M.J. Brown

Division of Civil Engineering, University of Dundee, Dundee, Scotland, UK

ABSTRACT

There is a need for an improved understanding of rate effects over a wide range of strain rates in order to improve the modelling and analysis of installation effects. Using triaxial testing on reconstituted kaolin over a wide strain rate range this paper examines the impact of strain rate on the aspects of soil response which are important in the analysis of installation effects. It is demonstrated that shear strength, small strain stiffness and the elastic shear strain threshold are rate dependent. The implications of this for a common stiffness degradation model are analysed and an improved model proposed.

REFERENCES

Bea, R.G. 1982. Soil strain rate effects on axial pile capacity. *Proc. 2nd Int. Conf. on Numerical Methods in Offshore Eng*: 107–132.

Brown, M. 2009. Recommendations for Statnamic use and interpretation of piles installed in clay. *Rapid Load Testing on Piles*: 23–36. London: Taylor & Francis.

Brown, M.J. & Powell, J.J.M. 2013. Comparison of rapid load test analysis techniques in clay soils. *ASCE Journal of Geotechnical & Geoenvironmental Engineering*. Available online 21/03/12.

Chow, S.H. & Airey, D.W. 2011. Rate effects in free falling penetrometer tests. *Proc. Int. Symp.on Deformation Characteristics of Geomaterials*, Seoul, 1–3 September 2011.

Krieg, S. & Goldscheider, M. 1998. Bodenviskotät und ihr Einfluß auf das Tragverhalten von Pfählen. *Bautechnik 75*: 806–820. Ernst und Sohn.

Kulhawy, F.H. & Mayne, P.W. 1990. Manual on estimating soil properties for foundation design. *Report EL-6800*, Electric Power Research Institute, Pala Alto.

Lehane, B.N., O'Loughlin, C.D., Gaudin, C. & Randolph, M.F. 2009. Rate effect on penetrometer resistance in kaolin. *Geotechnique* 59: 41–52.

Lo Presti, D.C.F., Jamiolkowski, M., Pallara, O. & Cavallaro, A. 1996. Rate and creep effect on the stiffness of soils. *ASCE GSP* 61: 166–180.

Mukabi, J.N. & Tatsuoka, F. 1999. Influence of reconsolidation stress history and strain rate on the behavior of kaolin over a wide range of strain. *Geotechnics for developing Africa*: 365–377. Rotterdam: Balkema.

Plaxis bv. 2011. *Material models manual 2011*. Delft: Plaxis.

Quinn, T.A.C. & Brown, M.J. 2011. Effect of strain rate on isotropically consolidated kaolin over a wide range of strain rates in the triaxial apparatus. *Proc. Int. Symp.on Deformation Characteristics of Geomaterials*, Seoul, 1–3 September 2011.

Randolph, M.F. & Hope, S. 2004. Effect of cone velocity on cone resistance and excess pore pressures. *Proc. Int. Symp. Eng. Practice and Performance of Soft Deposits*: 147–152.

Rowe, P.W. & Barden, L. 1964. Importance of free ends in triaxial testing. *ASCE journal of the soil mechanics and foundations division* 90 (SM1): 1–27.

Shibuya, S., Mitachi, T., Hosomi, A. & Hwang, S.C. 1996. Strain rate effects on stress-strain behavior as observed in monotonic and cyclic triaxial tests. *ASCE GSP: Measuring and modelling time dependent soil behavior* 61: 214–227.

Steenfelt, J.S. 1993. Sliding resistance for foundations on clay till. *Predictive soil mechanics*: 664–684. London: Thomas Telford.

Installation Effects in Geotechnical Engineering – Hicks et al. (eds)
© 2013 Taylor & Francis Group, London, ISBN 978-1-138-00041-4

Rate dependent shear strength of silt at low stresses

S. te Slaa & J. Dijkstra
Delft University of Technology, Delft, The Netherlands

ABSTRACT

The penetration resistance of freshly deposited silt is measured. Therefore, a miniature ball-cone is designed with high accuracy at the applied scale. This paper presents the operation characteristics of the newly designed ball-cone and presents the first results. The first results indicate that subtle differences in penetration resistance are resolved.

REFERENCES

Chung, S., Randolph, M. & Schneider, J. 2006. Effect of Penetration Rate on Penetrometer Resistance in Clay. Journal of Geotechnical and Geoenvironmental Engineering, 132(9): 1188–1196.

Jacobs, W. 2011. Sand-mud erosion from a soil mechanical perspective, Ph.D. thesis, Delft University of Technology.

Low, H.E., Randolph, M.F., Lunne, T., Andersen, K.H. & Sjursen, M.A. 2011. Effect of soil characteristics on relative values of piezocone, T-bar and ball penetration resitances. Géotechnique, 61(8): 13.

Randolph, M.F. & Hope, S. 2004. Effect of cone velocity on cone resistance and excess pore pressures, Int. Symp. on Engineering Practice and Performance of Soft Deposits, Osaka, Japan, pp. 147–152.

Randolph, M.F. & House, A.R. 2001. The complementary roles of physical and computational modelling. IJPMG—International Journal of Physical Modelling in Geotechnics, 1: 01–08.

Randolph, M.F., Martin, C.M. & Hu, Y. 2000. Limiting resistance of a spherical penetrometer in cohesive material. Géotechnique, 50(5): 573–582.

Roberts, J., Jepsen, R. & Gotthard, D. 1998. Effects of particle size and bulk density on erosion of quartz particles. Journal of Hydraulic Engineering, 124: 1261.

Winterwerp, J. & Van Kesteren, W. 2004. Introduction to the physics of cohesive sediment in the marine environment. Elsevier Science Ltd.

Winterwerp, J.C., van Kesteren, W.G.M., van Prooijen, B. & Jacobs, W. 2012. A conceptual framework for shear-flow induced erosion of soft cohesive sediment beds. Journal of Geophysical Research—Oceans.

Control of excess pore pressure development during pile installations in soft sensitive clay

T. Tefera, G. Tvedt & F. Oset
Norwegian Public Roads Administration, Norway

ABSTRACT

During the construction of the 148 m long Øvre Sund Bridge on soft sensitive clay in Drammen, in the south-eastern part of Norway, the stability of the river banks was a challenge. The installation of displacement piles strongly compresses the adjoining soils and leads to build up of excess pore water pressure. This temporary build up of excess pore water pressure, coupled with the sensitivity of the clay soil, causes the soil to lose a good fraction of its shear strength in the short term. Dissipation of the excess pore pressure generated during pile installations may allow pore pressure to rise in the vicinity of the river bank slope, thus leading to failure of the slope, even where the soil has not been remoulded.

A strict construction control mechanism during the construction of the Øvre Sund Bridge in Drammen was planned. The follow up of the development of excess pore pressure during piling activity was one of the measures taken. Based on stability analyses of the slope along the river banks a criterion for excess pore pressure margins due to pile installations was established. This paper discusses the result of this strict pore pressure control for the piling activity. The result shows a systematic monitoring of the development of excess pore pressure during pile installations in sensitive clay helps to observe the safety margins of the slopes continuously during pile installation, saves waiting time due to excess pore water pressure and avoids extra cost of the project due to the build up of excess pore pressure related to pile installations.

REFERENCES

Aas, G. 1975. Skred som følge av peleramming i bløt leire. NGI Publication 110. 49–54 (In Norwegian).

Eigenbord, K.D. & Issigonis, T. 1996. Pore-water pressures in soft to firm clay during driving of piles into underlying dense sand, Can. Geotech. J. 33, 209–218.

Fleming, K., Weltman, A., Randolph, M. & Elson, K. 2008. Piling Engineering, Taylor & Francis, New York.

Johansen, S. & Finstad, J.A. 2009. Øvre Sund bru alarmgren-se poretrykksmålere Grønland, RIG 026-Rev A. (In Norwegian).

Kirkebø, S. 2006. Forslag til sikkerhetsfilosofi, RIG 001. (In Norwegian).

Nr. 115513–4, Rv 283 HP: 400, Parsell Øvre Sund bru, Geo-teknisk rapport, 1. februar 2007 (In Norwegian).

Tvedt, G. & Tefera, T. 2009. Øvre Sund bru erfaringer med poretrykk og bæreevne under peleramming på Grøn-land, Notat. (In Norwegian).

Tefera, T. Tvedt, G. & Oset, F. 2011. Excess pore pressure during pile driving in soft sensitive clay. 15th ECSMG, Athens. 1285–1290.

Seabed pipelines: The influence of installation effects

D.J. White

Shell EMI Offshore Engineering, University of Western Australia, Australia

ABSTRACT

Seabed pipelines exemplify the type of geotechnical challenge addressed by the GEO INSTALL project. Pipeline laying involves complex soil-structure interaction. The seabed undergoes large deformations that are accompanied by changes in strength. Pipeline design requires assessment of the available pipe-soil interaction forces after installation, and the potential mobility and scour of the surrounding soil. These all depend on the installation process. This paper summarises recent research into the geotechnical aspects of pipeline installation, including numerical and physical modelling as well as field observations.

REFERENCES

An, H., Luo, C., Cheng, L. White, D. & Brown T. 2011. Introduction to the O-tube. *Proc. Sixth International Conference on Asian and Pacific Coasts.* Hong Kong, China.

Aubeny, C.P., Shi, H. & Murff, J.D. 2005. Collapse load for a cylinder embedded in trench in cohesive soil. *International Journal of Geomechanics,* 5(4): 320–325.

Bruton, D., Carr, M. & White, D.J. 2007. The influence of pipe-soil interaction on lateral buckling and walking of pipelines: the SAFEBUCK JIP. *Proc. 6th Int. Conf. on Offshore Site Investigation and Geotechnics,* London. 133–150.

Bruton, D.A.S., White, D.J., Carr, M.C. & Cheuk, C.Y. 2008. Pipe-soil interaction during lateral buckling and pipeline walking: the SAFEBUCK JIP. *Proc. Offshore Technology Conference,* Houston, USA. Paper OTC19589.

Biscontin, G. & Pestana, J.M. 2001. Influence of peripheral velocity on vane shear strength of an artificial clay. *ASTM Geotech. Test. J.* 24(4): 423–429.

Carr, M.C., Sinclair, F. & Bruton, D.A.S. 2006. Pipeline walking—understanding the field layout challenges, and analytical solutions developed for the SAFEBUCK JIP. *Proc. Offshore Tech. Conf.,* Houston, Paper OTC17945.

Casagrande, A. & Wilson, S.D. 1951. Effect of rate of loading on the strength of clays and shales at constant water content. *Géotechnique* 2(3): 251–263.

Chatterjee, S., Randolph, M.F., White, D.J. & Wang, D. 2010. Large deformation finite element analysis of vertical penetration of pipelines into the seabed. *Proc. 2nd Int. Symp. on Frontiers in Offshore Geotechnics.* Perth. 785–790.

Chatterjee, S., Randolph, M.F. & White, D.J. 2012a. The effects of penetration rate and strain softening on the vertical penetration resistance of seabed pipelines. *Géotechnique,* 62(7): 573–582.

Chatterjee, S., Yan, Y., Randolph, M.F. & White, D.J. 2012b. Elastoplastic consolidation beneath shallowly embedded offshore pipelines. *Géotechnique Letters,* 2(2): 73–79.

Chatterjee, S., White, D.J. & Randolph, M.F. 2013. Coupled consolidation analysis of pipe-soil interactions. *Canadian Geotechnical Journal,* in review.

Cheuk, C.Y. & White, D.J. 2011. Modelling the dynamic embedment of seabed pipelines. *Géotechnique.* 61(1): 39–57.

Cheng, L. White, D.J., Brown, T. An, H. & Luo, C. 2010. *O-tube Pipeline Stability Project, Development of the Large O-tube Facility, Commissioning Report.* UWA Report GEO: 10500 v2. 115 pp.

De Catania, S., Breen, J., Gaudin, C. & White, D.J., 2010. Development of multiple-axis actuator control system *Proc. Int. Conf. on Physical Modelling in Geotechnics.* Zurich, Switzerland, 325–330.

Dingle, H.R.C., White, D.J., & Gaudin, C. 2008. Mechanisms of pipe embedment and lateral breakout on soft clay. *Canadian Geotechnical Journal,* 45(5): 636–652.

Einav, I., & Randolph, M.F. 2005. Combining upper bound and strain path methods for evaluating penetration resistance. *International Journal for Numerical Methods in Engineering,* 63(14): 1991–2016.

Gaudin, C., White, D.J., Boylan, N., Breen, J., Brown, T. & De Catania, S. 2010. Development of a miniature high speed wireless data acquisition system for geotechnical centrifuges *Proc. Int. Conf. on Physical Modelling in Geotechnics.* Zurich, Switzerland, 229–234.

Gourvenec, S.M. & White, D.J. 2010. Elastic solutions for consolidation around seabed pipelines. *Proc. Offshore Technology Conference,* Houston. Paper 20554.

Gourvenec, S.M. & White, D.J. 2013. Softening and consolidation around seabed pipelines: centrifuge modelling. *In review.*

Graham, J., Crooks, J.H.A. & Bell, A.L. 1983. Time effects on the stress-strain behaviour of natural soft clays. *Géotechnique* 33(3): 327–340.

Green, A.P. 1954. The plastic yielding of metal junctions due to combined shear and pressure, *J. Mechanics and Physics of Solids,* Vol. 2, pp. 197–211.

Hill, A., White, D.J., Bruton, D.A.S., Langford, T., Meyer, V., Jewell, R. & Ballard J-C. 2012. A new framework for axial pipe-soil interaction illustrated by a range of marine clay datasets. *Proc. Int. Conf. on Offshore Site Investigation and Geotechnics.* SUT, London.

Hu, Y. & Randolph, M.F. 1998a. A practical numerical approach for large deformation problems in soil. *Int. J. Numerical and Analytical Meth. Geomech.* 22(5): 327–350.

Hu, Y. & Randolph, M.F. 1998b. H-adaptive FE analysis of elastoplastic non-homogeneous soil with large deformation. *Computers and Geotechnics*. 23(1–2): 61–83.

Jardine, R.J., Chow, F.C., Overy, R. & Standing, J.R. 2010. *ICP design methods for diven piles in sands and clays*. Thomas Telford.

Jas, E., O'Brien, D., Fricke, R., Gillen, A., Cheng, L. & White, D.J. 2013. Pipeline stability revisited. *Journal of Pipeline Engineering*. in press.

Jayson, D., Delaporte, P., Albert, J-P, Prevost, M.E., Bruton, D. & Sinclair, F. 2008. Greater Plutonio project—Subsea flowline design and performance. *Offshore Pipeline Technology Conf.*, Amsterdam.

Krost, K., Gourvenec, S.M. & White, D.J. 2011. Consolidation around partially-embedded submarine pipelines. *Géotechnique*, 61(2): 167–173.

Lenci, S. & Callegari., M. 2005. Simple analytical models for the J-lay problem. *Acta Mechanica*, 178: 23–39.

Low, H.E., Randolph, M.F., DeJong, J.T. & Yafrate, N.J. 2008. Variable rate full-flow penetration tests in intact and remouldedsoil. *Proc. 3rd Int. Conf. on Geotech. Geophys. Site Characterization*, Taipei, Taiwan 1087–1092.

Lund, K.M. 2000. Effect of increase in pipeline soil penetration from installation. *Proc. of ETCE/OMAE2000 Joint Conference; Energy of the New Millennium* Paper OMAE2000/PIPE-5047.

Lunne, T., Berre, T., Andersen, K.H., Strandvik, S. & Sjursen, M. 2006. Effects of sample disturbance and consolidation procedures on measured shear strength of soft marine Norwegian clays. *Can. Geotech. J.* 43(7): 726–750.

Lunne, T. & Andersen, K.H. 2007. Soft clay shear strength parameters for deepwater geotechnical design. *Proc. 6th Int. Conf. Offshore Site Investigation and Geotechnics: Confronting New Challenges and Sharing Knowledge*, SUT, London, Vol. 1: 151–176.

Martin, C.M. & White, D.J. 2012. Limit analysis of the undrained capacity of offshore pipelines. *Géotechnique*, 62(9): 847–863

Merifield, R.S., White, D.J. & Randolph, M.F. 2008a. Analysis of the undrained breakout resistance of partially embedded pipelines. *Géotechnique* 58(6): 461–470.

Merifield R., White D.J. & Randolph M.F. 2009. The effect of surface heave on the response of partially-embedded pipelines on clay. *ASCE J. Geotechnical & Geoenvironmental Engineering*, 135(6): 819–829.

Murff, J.D., Wagner, D.A. & Randolph, M.F. 1989. Pipe penetration in cohesive soil. *Géotechnique* 39(2): 213–229.

Orcina. 2008. *OrcaFlex software*, Orcina Ltd, Ulverston, UK.

Palmer, A. 2009. Touchdown indentation of the seabed, *Applied Ocean Research*, Vol. 30, No. 3. pp. 235–238.

Pesce, C.P., Aranha, J.A.P. & Martins, C.A. 1998. The soil rigidity effect in the touchdown boundary layer of a catenary riser: Static problem, *Proc. 8th Int. Offshore and Polar Engineering Conf.*, Montreal, pp. 207–213.

Randolph, M.F., Carter, J.P., & Wroth, C.P. 1979. Driven piles in clay—the effects of installation and subsequent consolidation. *Géotechnique* Vol. 29, No. 4, 361–393.

Randolph, M.F. & White, D.J. 2008a. Upper bound yield envelopes for pipelines at shallow embedment in clay. *Géotechnique*, 58(4): 297–301.

Randolph, M.F. & White, D.J. 2008b. Pipeline embedment in deep water: processes and quantitative assessment. *Proc. Offshore Technology Conference*, Houston, USA. Paper OTC19128-PP.

Randolph, M.F. & Gourvenec, S.M. 2011. *Offshore Geotechnical Engineering*. ISBN 978-0415477444. 560 pages. Taylor & Francis.

Wang, D., White, D.J. & Randolph, M.F. 2009. Numerical simulations of dynamic embedment during pipelaying on soft clay. *Proc. Conf. on Offshore Mechanics and Arctic Engineering*, Honolulu. Paper OMAE2009-79199.

Wang, D, White, D.J. & Randolph, M.F. 2010. Large deformation finite element analysis of pipe penetration and large-amplitude lateral displacement. *Canadian Geotechnical Journal*, 47(8): 842–856.

Westgate, Z.J., White, D.J. and Randolph, M.F. 2009. Video observations of dynamic embedment during pipelaying on soft clay. *Proc. Conf. on Offshore Mechanics and Arctic Engineering*, Honolulu. Paper OMAE2009-79814.

Westgate, Z.J., Randolph, M.F., White D.J. & Li, S. 2010a. The effect of seastate on as-laid pipeline embedment: a case study, *Applied Ocean Research*, Vol. 32, pp. 321–331.

Westgate, Z.J., White, D.J., Randolph, M.F. & Brunning, P. 2010b. Pipeline laying and embedment in soft fine-grained soils: field observations and numerical simulations, *Proc. Offshore Tech. Conf.*, Paper OTC20407.

Westgate, Z., White, D.J. & Randolph, M.F. 2012. Field observations of pipeline embedment in carbonate sediments. *Géotechnique*, 62(9): 787–798.

Westgate Z.J., White D.J. & Randolph M.F. 2013. Modelling the embedment process during offshore pipe laying on fine-grained soils. *Canadian Geotechnical Journal*, in press.

White, D.J. & Hodder, M. 2010. A simple model for the effect on soil strength of remoulding and reconsolidation. *Canadian Geotechnical Journal*, 47: 821–826.

Zeitoun H.O., Tornes K., Cumming, G., & Branković, M. 2008. Pipeline stability—state of the art. *Proc. 27th Int. Conf. on Offshore Mechanics and Arctic Engineering, OMAE2008*, Estoril, Portugal.

Zhou, H. & Randolph, M.F. 2007. Computational techniques and shear band development for cylindrical and spherical penetrometers in strain-softening clay. *Int. J. Geomechanics* 7(4): 287–295.

Zhou, H. & Randolph, M.F. 2009. Penetration resistance of cylindrical and spherical penetrometers in rate-dependent and strain-softening clay. *Géotechnique*. 59(2): 79–86.

Zienkiewicz, O.C. & Zhu, J.Z. 1992. The superconvergent patch recovery and a posterior error estimates. Part 1: The recovery technique. *International Journal of Numerical Methods in Engineering*, 33(7): 1331–1364.

Offshore construction and foundations

LDFE analysis of installation effects for offshore anchors and foundations

L. Andresen & H.D.V. Khoa

Norwegian Geotechnical Institute, Oslo, Norway

ABSTRACT

During installation of offshore anchors and foundations the soil undergoes large deformations. Effects of installation such as penetration resistance, punch-through, changes in contact stresses and strengths should be accounted for in the design and planning. Typically such effects have been assessed by semi empirical methods and engineering judgement skills obtained from field and model testing. Numerical methods for modelling large deformation and penetration problems in geomaterials are now available, and although research into new developments and improvements is still ongoing, these methods can already be used for some design problems. This paper presents and briefly reviews two modelling approaches: the Arbitrary Lagrangian-Eulerian method and the Coupled Eulerian-Lagrangian method of Abaqus/Explicit. The CEL method is found to be the most promising of the two and this method is used to model the penetration of a spudcan foundation. Excellent agreement is obtained with the penetration resistance and failure mechanism measured in a centrifuge test.

REFERENCES

Cudmani, R. & Sturm, H. 2006. An investigation of the tip resistance in granular and soft soils during static, alternating and dynamic penetration. In H. Gonin, A. Holeyman, and F. Rocher-Lacoste (Eds.), *TransVib 2006: International Symposium on vibratory pile driving and deep soil compaction*: 221–231.

Hossain, M.S. & Randolph, M.F. 2010a. Deep-penetrating spudcan foundations on layered clays: Centrifuge tests. *Géotechnique*. 60(3): 157–170.

Hossain, M.S. & Randolph, M.F. 2010b. Deep-penetrating spudcan foundations on layered clays: Numerical analysis. *Géotechnique*, 60(3): 171–184.

Hu, Y. & Randolph, M.F. 1998. A practical numerical approach for large deformation problems in soil. *Int. J. Numer. Anal. Methods Geomech.*, 22(5): 327–350.

Liyanapathirana, D.S. 2009. Arbitrary Lagrangian Eulerian based finite element analysis of cone penetration in soft clay. *Computers and Geotechnics* 36: 851–860.

Menzies, D. & Roper, R. 2008. Comparison of jackup rig spudcan penetration methods in clay. *Proc. Offshore Technology Conference OTC-19545.*

Qiu, G. & Henke, S. 2011. Controlled installation of spudcan foundations on loose sand overlying weak clay. *Marine Structures* 24(4): 528–550.

Qiu, G., Henke, S. & Grabe, J. 2011. Applications of a Coupled Eulerian-Lagrangian approach on geotechnical problems involving large deformations. *Computers and Geotechnics* 38(1): 30–39.

Randolph, M.F., Wang, D., Zhou, H., Hossain, M.S. & Hu, Y. 2008. Large deformation finite element analysis for offshore applications. *Proc., 12th Int. Conf. of Int. Association for Computer Methods and Advances in Geomechanics*: 3307–3318. Goa, India.

Tho, K.K., Leung, C.F., Chow, Y.K. & Swaddiwudhipong, S. 2012. Eulerian finite-element technique for analysis of jack-up spudcan penetration. *Int. J. Geomech.* 12(1): 64–73.

Wang, C.X. & Carter, J.P. 2002. Deep penetration of strip and circular footings into layered clays. *Int. J. Geomech.* 2(2): 205–232.

Investigation into the effect of pile installation on cyclic lateral capacity of monopiles

T. de Blaeij & J. Dijkstra

Delft University of Technology, Delft, The Netherlands

ABSTRACT

Monopiles are being used more extensively for offshore wind turbine foundation. Research into the behaviour of these large open-ended piles generally do not examine the installation effect on lateral capacity. This paper presents the results of model pile load tests on monopiles in the geotechnical centrifuge. The effect of monotonic pile installation on the subsequent cyclic lateral capacity were investigated. In order to study these effects, a novel actuator was developed which simultaneously allows to install the piles in-flight and is able to load the piles laterally without interrupting the test. A series of initial tests on open-ended monopiles, which show the effectiveness of the setup, are presented. These initial tests investigated the effect of 1·g and N·g pile installation on the subsequent lateral capacity in two way cycling for two different initial densities. The preliminary results indicate that N·g installation has a small positive effect, which decays with the number of cycles, on the lateral capacity of short stiff monopiles. Hence, for open-ended model piles the installation effects are negligible. At the present stage, given the accuracy of the experimental setup, no conclusive results for the initial stiffness response were gathered. However, tentatively, the conclusions for the lateral capacity also hold for the initial stiffness. The latter needs to be confirmed in future experiments, where the setup is optimized for the measurement of the small strain response of the monopile and a large number of loading cycles. These conclusions are for tests on open-ended model piles only. If for similitude a close-ended model pile will be used the pile installation stage could be omitted in order to better approach the negligible installation effects as presented in this paper.

REFERENCES

Achmus, M., Abdel-Rahman, K. & Kuo, Y. 2007. Numerical modelling of large diameter steel piles under monotonic and cyclic horizontal loading. In *Tenth International Symposium on Numerical Models in Geomechanics*, pp. 453–459. Taylor & Francis London.

Alderlieste, E. 2011. Experimental modelling of lateral loads on large diameter mono-pile foundations in sand. Msc thesis, Delft University of Technology.

Alderlieste, E.A., Dijkstra, J., & van Tol, A.F. 2011. Experimental investigation into pile diameter effects of laterally loaded mono-piles. *ASME Conference Proceedings* 2011(44397), 985–990.

API 2007. *Recommended Practice for Planning, Design and Constructing Fixed Offshore Platforms—Working Stress Design.* American Petroleum Institute. ERRATA AND SUPPLEMENT 3, AUGUST 2007.

Barton, Y., Fin,W., Pary, R., & Ikuo, T. 1983. Lateral pile response and p-y curves from centrifuge tests. *Offshore Technology Conference Paper number OTC 4502.*

Bienen, B., Dührkop, J., Grabe, J., Randolph, M., & White, D. 2011. Response of piles with wings to monotonic and cyclic lateral loading in sand. *Journal of Geotechnical and Geoenvironmental Engineering* 138(3), 364–375.

Brant, L. & Ling, H. 2007. Centrifuge modeling of piles subjected to lateral loads. In H.I. Ling, L. Callisto, D. Leshchinsky, J. Koseki, and G.M.L. Gladwell (Eds.), *Soil Stress-Strain Behavior: Measurement, Modeling and Analysis*, Volume 146 of Solid Mechanics and Its Applications, pp. 895–907. Springer Netherlands. 10.1007/978-1-4020-6146-268.

Byrne, B., Leblanc, C., & Houlsby, G. 2010. Response of stiff piles in sand to long-term cyclic lateral loading. *Géotechnique* 60(2), 79–90.

Clauss, G., Lehmann, E., & Östergaard, C. 1988. *Meerestechnische Konstruktionen.* Springer.

Craig,W. 1985. Installation studies for model piles. *Publication of: Balkema (AA).*

Cuéllar, P., Georgi, S., Baeßler, M., & Rücker, W. 2012. On the quasi-static granular convective flow and sand densification around pile foundations under cyclic lateral loading. *Granular Matter*, 11–25.

Dijkstra, J. 2009. On the Modelling of Pile Installation. Ph. D. thesis, Technische Universiteit Delft. ISBN: 9789085704324. DNV 2011. Offshore standard dnv-os-j101—design of offshore wind turbine structures.

Dührkop, J. & Grabe, J. 2008. Monopilegründungen von offshore-windenergieanlagen–zum einfluss einer veränderlichen zyklischen lastangriffsrichtung. *Bautechnik* 85(5), 317–321.

Dyson, G. & Randolph, M. 2001. Monotonic lateral loading of piles in calcareous sand. *Journal of Geotechnical and Geo-environmental Engineering* 127, 346–352.

Garnier, J. & König, D. 1998. Scale effects in piles and nails loading tests in sand. In *Centrifuge 98.* GL 2005. Guideline for the certification of offshore wind turbines.

Klinkvort, R., Leth, C. & Hededal, O. 2010. Centrifuge modelling of a laterally cyclic loaded pile. *Physical Modelling in Geotechnics*, 959–964.

Li, Z., Haigh, S.K., & Bolton, M.D. 2010. Centrifuge modelling of mono-pile under cyclic lateral loads. *7th International Conference on Physical Modelling in Geotechnics* 2, 965–970.

Murchison, J. & O'Neill, M. 1983. An Evaluation of Py Relationships in Sands. Report (American Petroleum Institute). University of Houston-University Park.

Oldham, D. 1985. Experiments with lateral loading of single piles in sand. Publication of: Balkema (AA).

Paikowsky, S., Player, C., & Connors, P. 1995. A dual interface apparatus for testing unrestricted friction of soil along solid surfaces. *ASTM geotechnical testing journal* 18(2), 168–193.

Peng, J., Clarke, B. & Rouainia, M. 2011. Increasing the resistance of piles subject to cyclic lateral loading. *Journal of Geotechnical and Geoenvironmental Engineering* 137(10), 977–982.

Reese, L., Cox, W. & Koop, F. 1974. Analysis of laterally loaded piles in sand. *Offshore Technology Conference OTC 2080*, 473–486.

Rosquoet, F., Thorel, L., Garnier, J. & Canepa, Y. 2007. Lateral cyclic loading of sand-installed piles. *Soils and foundations* 47(5), 821–832.

Vattenfall 2008. Kriegers flak offshore wind farm-design basis foundations. Technical report, Vattenfall Vindkracft AB.

White, D.J. & Lehane, B. 2004. Friction fatigue on displacement piles in sand. *Géotechnique* 54, 645–658.

Installation Effects in Geotechnical Engineering – Hicks et al. (eds)
© 2013 Taylor & Francis Group, London, ISBN 978-1-138-00041-4

Development of a coupled FEM-MPM approach to model a 3D membrane with an application of releasing geocontainer from barge

F. Hamad & C. Moormann
Institute of Geotechnics, Stuttgart, Germany

P.A. Vermeer
Deltares, Delft, The Netherlands
Institute of Geotechnics, Stuttgart, Germany

ABSTRACT

In a wide spectrum of geotechnical applications, materials undergo large deformations and/or large displacements. On modeling these problems with a Lagrangian finite element method, the mesh can become too distorted and re-meshing is essential. In the past decades, considerable efforts have been made to adopt what is called mesh free methods to mitigate the problems related to mesh distortion. One of these methods is the Material Point Method (MPM) that represents the continuum field as Lagrangian material points (particles), which can move through the fixed background of an Eulerian mesh. The objective of this paper is to formulate and validate a coupled FEM-MPM approach for the numerical simulation of large deformation of a membrane containing soil e.g. dumping of a geocontainer. In this approach the membrane is discretised by a surface mesh with accurate computation of the membrane stresses. This membrane mesh is free to move through a 3D mesh of non-structured tetrahedral elements. Furthermore, the proposed approach is applied successfully to model geotextile of a geocontainer being released from a splitbarge. Frictional contact is defined between the geotextile and the barge. On the other side, rough contact is assumed between the geotextile and the Mohr-Coulomb soil type inside.

REFERENCES

Bathe, K. 1996. *Finite element procedures*, Volume 2. Prentice hall Englewood Cliffs, NJ.

Bezuijen, A., R.R., S. & Klein Breteler, M., Berendesen E., P.K. 2002. Field tests on geocontainers. In *Proceedings of 7th International conference on Geosythetics*, Nice.

Burgess, D., Sulsky, D. & Brackbill, J. 1992. Mass matrix formulation of the flip particle-in-cell method. *Journal of Computational Physics* 103(1), 1–15.

De Groot, M. & Bezuijen, A. 2000. Forces in eocontainer geotextile during dumping from barge. In *Proceedings of 2nd EuroGeo2*, pp. 623–627.

Fernandes, I. 2004. Exploring quadratic shape functions in material point method. Masters thesis, Department of Mechanical Engineering, The University of Utah.

Jassim, I., Hamad, F., & Vermeer, P. 2011, 27–29 April. Dynamic material point method with applications in geomechanics. In *2nd International Symposium on Computational Geomechanics (COMGEO II)*, Cavtat-Dubrovnik, Croatia.

Pilarczyk, K. 2000. Geosynthetics and geosystems in hydraulic and coastal engineering. Taylor & Francis.

Sulsky, D. & Schreyer, H. 1993. A particle method with large rotations applied to the penetration of history dependent materials. *Advances in Numerical Simulation Techniques for Penetration and Perforation of Solids, ASME, AMD AMD-Vol 171*, 95–102.

Sulsky, D., Zhou, S. & Schreyer, H. 1995. Application of a particle-in-cell method to solid mechanics. *Computer Physics Communications* 87(1), 236–252.

Wieckowski, Z., Youn, S. & Yeon, J. 1999. A particlein-cell solution to the silo discharging problem. *International journal for numerical methods in engineering* 45(9), 1203–1226.

York, A.R. 1997. The development of modifications to the material point method for the simulation of thin membranes, compressible fluids, and their interactions. Ph.D. thesis, The University of New Mexico.

Zhang, D., Ma, X. & Giguere, P. 2011. Material point-method enhanced by modified gradient of shape function. *Journal of Computational Physics* 230, 6379–6398.

Leg penetration assessments for self-elevating tubular leg units in sand over clay conditions

D.A. Kort
Norwegian Geotechnical Institute, Oslo, Norway

S. Raymackers
GeoSea NV, Zwijndrecht, Belgium

H. Hofstede
Gusto MSC BV, Schiedam, The Netherlands

V. Meyer
Norwegian Geotechnical Institute, Oslo, Norway

ABSTRACT

In a wide spectrum of geotechnical applications, materials undergo large deformations and/or large displacements. On modeling these problems with a Lagrangian finite element method, the mesh can become too distorted and re-meshing is essential. In the past decades, considerable efforts have been made to adopt what is called meshfree methods to mitigate the problems related to mesh distortion. One of these methods is the Material Point Method (MPM) that represents the continuum field as Lagrangian material points (particles), which can move through the fixed background of an Eulerian mesh. The objective of this paper is to formulate and validate a coupled FEM-MPM approach for the numerical simulation of large deformation of a membrane containing soil e.g. dumping of a geocontainer. In this approach the membrane is discretised by a surface mesh with accurate computation of the membrane stresses. This membrane mesh is free to move through a 3D mesh of non-structured tetrahedral elements. Furthermore, the proposed approach is applied successfully to model geotextile of a geocontainer being released from a split barge. Frictional contact is defined between the geotextile and the barge. On the other side, rough contact is assumed between the geotextile and the Mohr-Coulomb soil type inside.

REFERENCES

Bathe, K. 1996. *Finite element procedures*, Volume 2. Prentice hall Englewood Cliffs, NJ.

Bezuijen, A., S.R.R. & P.K. Klein Breteler M., Berendesen E. 2002. Field tests on geocontainers. In *Proceedings of 7th Internationalconference on Geosythetics*, Nice.

Burgess, D., Sulsky, D. & Brackbill, J. 1992. Mass matrix formulation of the flip particlein-cell method. *Journal of ComputationalPhysics 103*(1), 1–15.

De Groot, M. & Bezuijen, A. 2000. Forces in geocontainer geotextile during dumping frombarge. In *Proceedings of 2nd EuroGeo 2*, pp. 623–627.

Fernandes, I. 2004. Exploring quadratic shape functions in material point method. Masters thesis, Department of Mechanical Engineering, The University of Utah.

Jassim, I., Hamad, F. & Vermeer, P. 2011. Dynamic material point method with applications in geomechanics. In *2nd International Symposium on Computational Geomechanics (COMGEO II)*, Cavtat-Dubrovnik, Croatia.

Pilarczyk, K. 2000. *Geosynthetics and geosystemsin hydraulic and coastal engineering*. Taylor & Francis.

Sulsky, D. & Schreyer, H. 1993. A particle method with large rotations applied to the penetration of history-dependent materials. *Advances in Numerical Simulation Techniques for Penetration and Perforation of Solids, ASME, AMD AMD-Vol 171*, 95–102.

Sulsky, D., Zhou, S. & Schreyer, H. 1995. Application of a particle-in-cell method to solid mechanics. *Computer Physics Communications 87*(1), 236–252.

Wieckowski, Z., Youn, S. & Yeon, J. 1999. A problem. *International journal for numerical methods in engineering 45*(9), 1203–1226.

York, A.R. 1997. *The development of modifications to the material point method for the simulation of thin membranes, compressible fluids, and their interactions*. Ph.D. thesis, The University of New Mexico.

Zhang, D., Ma, X. & Giguere, P. 2011. Material point method enhanced by modified gradient of shape function. *Journal of Computational Physics 230*, 6379–6398.

Installation Effects in Geotechnical Engineering – Hicks et al. (eds)
© 2013 Taylor & Francis Group, London, ISBN 978-1-138-00041-4

Investigating the scales of fluctuation of an artificial sand island

M. Lloret-Cabot

Centre for Geotechnical and Materials Modelling, The University of Newcastle, Newcastle, Australia
Department of Geoscience and Engineering, Delft University of Technology, Delft, The Netherlands

M.A. Hicks & J.D. Nuttall

Department of Geoscience and Engineering, Delft University of Technology, Delft, The Netherlands

ABSTRACT

Leg penetration assessments were performed for two self-elevating units with tubular legs operating at 21 locations within an offshore wind farm. The assessments were performed on the basis of available CPT data, experience-based soil parameters and the calculation methods recommended by SNAME T&RB 5–5A. In general the soil conditions are dense sand over stiff clay with strong variations of the sand layer thickness across the wind farm. Initially at some locations the assessment indicated a significant risk of punch-through, which could potentially lead to problems with insufficient leg length and create difficulties for leg extraction after installation of the wind turbines. To reduce the uncertainties of the leg penetration assessments, preload trials were performed at three critical locations. The three preload trials provided a useful basis to revise the input soil parameters and thereby enhance the leg penetration predictions. No indications of large penetrations or rapid penetrations were encountered during the offshore field campaign.

REFERENCES

Andersen, K.H. & Schjetne, K. 2012. Data base of friction angles and consolidation characteristics. Accepted for publication in *ASCE, Journal of Geotechnical and Geoenvironmental Engineering*.

Baglioni, V.P., Chow, G.S. & Endley, S.N. 1982. Jack-up rig foundation stability in stratified soil profiles. *Journal of Offshore Technology*, OTC 4409, pp. 363–383.

Baldi, G., Bellotti, R., Ghionna, V. Jamiolkowski, M. & Pasqualini, E. 1986. Interpretation of CPTs and CPTUs; 2nd part: drained penetration of sands. *Proceedings of the 4th International Geotechnical Seminar Singapore*: 143–56.

Bjerrum, L. 1973. Problems of Soil Mechanics and Construction on Soft Clays. State-of-the-Art report to Session IV, Proceedings 8th ICSMFE, Moscow. Vol. 3, pp. 111–159.

Dijkstra, J. 2009. *On the modelling of pile installation*. Dr. Thesis Delft University of Technology. Zutphen: Wöhrmann Print Service.

DNV 1992. DNV Classification Note 30.4. Det Norske Veritas.

Gregg 2007. Cone Penetration Test (CPT) Interpretation. http://www.greggdrilling.com/PDF_files/TechnicalMethodologyPDF/cptinterpretationsummary.pdf. Downloaded 15 December 2011.

ISO 19905-1 2012. Petroleum and natural gas industries. Site-specific assessment of mobile offshore units—Part 1: Jack-ups.

Lambe, T.W. & Whitman, R.V. 1979. *Soil Mechanics*. New York: John Wiley & Sons.

Lee, F.H. 2009. Investigation of potential punch-through failure on sands overlying clay soils. Ph.D. thesis, The University of Western Australia.

Lunne, T., Robertson, P.K. & Powell, J.J.M. 1997. *Cone Penetration Testing in Geotechnical Practice*. London: Blackie Academic & Professional.

Plaxis 2008. Plaxis Version 9.02. www.plaxis.nl.

Purwana, O.A., Quah, M. Foo, K.S. Nowak S. & Handidjaja, P. 2009. Leg Extraction/Pullout Resistance—Theoretical and Practical Perspectives. *Proceedings of the 12th International Conference The Jack-Up Platform Design, Construction & Operation, London*.

SNAME 2008. Technical and Research Bulletin 5–5A. *Guidelines for Site Specific Assessment of Mobile Jack-Up Units*. Society of Naval Architects and Marine Engineers, Jersey City, New Jersey. August 2008.

Schroeder, K., Andersen, K.H. & Jeanjean, P. 2006. Predicted and observed installation behavior of the Mad Dog anchors. *Proceedings of the Offshore Technology Conference, Houston*, OTC 17950.

Teh, K.L., Leung, C.F. Chow, Y.K. & Cassidy, M.J. 2010. Centrifuge model study of spudcan penetration in sand overlying clay. *Proceedings of the Offshore Technology Conference, Houston*, OTC 20060.

Soil improvement

Installation Effects in Geotechnical Engineering – Hicks et al. (eds)
© *2013 Taylor & Francis Group, London, ISBN 978-1-138-00041-4*

Volume averaging technique in numerical modelling of floating deep mixed columns in soft soils

P. Becker
Kempfert und Partner Geotechnik, Hamburg, Germany, previous University of Strathclyde, Glasgow, Scotland, UK

M. Karstunen
Chalmers University of Technology, Göteborg, Sweden
University of Strathclyde, Glasgow, Scotland, UK

ABSTRACT

Artificial sand islands were constructed in the Canadian Beaufort Sea for use as hydrocarbon exploration platforms in the 1970s and 1980s. For some of these islands, extensive Cone Penetration Test (CPT) data are available for characterising the hydraulically placed sand during and after the construction process. Tarsiut P-45 was the first island using the 'Molikpaq' concept, which consisted of a mobile arctic caisson system to provide the temporary structure for the exploitation. Two main sand fills were constructed: (a) a sandfill berm on which the caisson system was founded; and (b) the body of the island structure (island core). This paper presents an investigation of the variability of the sand in the berm in terms of the vertical and horizontal scales of fluctuation. This geo-statistical investigation is carried out using CPT data from the berm before and after the founding of the caisson system, and sets the basis for a preliminary discussion on the potential soil variability changes caused by the installation and infilling of the caisson structure when placed on the berm.

REFERENCES

Baecher, G.B. & Christian, J.T. 2003. *Reliability and statistics in geotechnical engineering.* John Wiley & Sons Inc.

Fenton, G.A. & Griffiths, D.V. 2008. Risk assessment in geotechnical engineering. John Wiley & Sons, New Jersey, USA.

Griffiths, D.V. & Fenton, G.A. 2001. Bearing capacity of spatially random soil: the undrained clay Prandtl problem revisited. *Géotechnique* 51(4): 351–359.

Hicks, M.A. & Onisiphorou, C. 2005. Stochastic evaluation of static liquefaction in a predominantly dilative sand fill. *Géotechnique* 55(2): 123–133.

Hicks, M.A. & Samy, K. 2002. Influence of heterogeneity on undrained clay slope stability. *Quarterly Journal of Engineering Geology and Hydrogeology* 35(1): 41–49.

Hicks, M.A. & Smith, I.M. 1988. Class A prediction of Arctic caisson performance. *Géotechnique* 38(4): 589–612.

Hicks, M.A. & Spencer, W.A. 2010. Influence of heterogeneity on the reliability and failure of a long 3D slope. *Computers and Geotechnics* 37(7–8): 948–955.

Lloret, M., Hicks, M.A. & Wong, S.Y. 2012 Soil characterisation of an artificial island accounting for soil heterogeneity. *GeoCongress 2012, R.D. Hryciw, A. Athanasopoulos-Zekkos & N. Yesiller (Eds.),* San Francisco, 2816–2825.

Lloret-Cabot, M., Hicks, M.A. & Eijnden, A.P. van den. 2012. Investigation of the reduction in uncertainty due to soil variability when conditioning a random field using Kriging. *Géotechnique letters* 2: 123–127.

Nuttall, J.D. 2011. Parallel implementation and application of the random finite element method. Ph.D. thesis, University of Manchester, UK.

Phoon, K-K. & Kulhawy, F.H. 1999. Characterization of geotechnical variability, *Canadian Geotechnical Journal* 36(4): 612–624.

Spencer, W.A. 2007. *Parallel stochastic and finite element modelling of clay slope stability in 3D.* Ph.D. thesis, University of Manchester, UK.

Uzielli, M., Vannucchi, G. & Phoon, K-K. 2005. Random field characterisation of stress-normalised cone penetration testing parameters. *Géotechnique* 55(1): 3–20.

Vanmarcke, E.H. 1977. Probabilistic modeling of soil profiles, *Journal of the Geotechnical Engineering Div. ASCE* 103(11): 1227–1246.

Vanmarcke, E.H. 1984. *Random Fields: Analysis and Synthesis.* The MIT Press, Cambridge, Massachusetts.

Wackernagel, H. 2003. *Multivariate geostatistics: An introduction with applications.* Springer, Germany.

Wickremesinghe, D.S. 1989. *Statistical characterization of soil profiles using in situ tests.* PhD thesis, University of British Columbia, Canada.

Wickremesinghe, D.S. & Campanella, R.G. 1993. Scale of fluctuation as a descriptor of soil variability. *Proc. Conf. Probabilistic Methods in Geotechnical Engineering,* Canberra, 233–239.

Wong, S.Y. 2004. *Stochastic characterisation and reliability of saturated soils.* Ph.D. thesis, University of Manchester, UK.

Installation Effects in Geotechnical Engineering – Hicks et al. (eds)
© 2013 Taylor & Francis Group, London, ISBN 978-1-138-00041-4

Comparison between theoretical procedures and field test results for the evaluation of installation effects of vibro-stone columns

E. Carvajal & G. Vukotić
Kellerterra S.L., Madrid, Spain

J. Castro
University of Cantabria, Santander, Spain

W. Wehr
Keller Holding GmbH, Offenbach, Germany

ABSTRACT

Several theoretical procedures to estimate the soil improvement produced by installation of vibro-stone columns are described. Particularly, finite element model and analytical solutions of a cylindrical cavity expansion were compared with results from an actual field test which was performed in silty sand and clayey soil treated with a column group. The results show that after dissipation of pore pressure the installation effects produce considerable improvement due to a large increase of the horizontal effective stress and due to densification process of sand. The load settlements response of the tested column group has been analyzed and compared with theoretical estimation of the improvement with and without installation effects, and with Priebe's analytical solution. It is observed that the column group installation effects have an important influence that should be evaluated with more advanced modeling or directly with in situ testing.

REFERENCES

Arnold, M., Herle, I. & Wehr, J. 2008. Comparison of vibrocompaction methods by numerical simulations. In Karstunen et al. (eds), *Geotechnics of Soft Soils: Focus on Ground Improvement*: University of Strathclyde, Glasgow.

Baguelin, F., Jezequel, J.F. & Shields, D.H. 1978. *The Pressuremeter and Foundation Engineering.* TransTech Publications.

Carter, J.P., Randolph, M.F. & Wroth, C.P. 1979. Stress and pore pressure changes in clay during and after the expansion of a cylindrical cavity. *International Journal for Numerical and Analytical Methods in Geomechanics.*

Castro, J. & Karstunem, M. 2010. Numerical simulations of stone column installation. *Canadian Geotechnical Journal 47*(19): 1127–1138.

Castro, J., Kamrat-Pietraszewska & D. Karstunem, M. 2012. Numerical modelling of stone column installation in Bothkennar clay. *International Symposium on Ground Improvement IS-GI Brussels; ISSMGE-TC 211.*

Egan, D., Scott, W. & McCabe, B. 2008. Installation effects of vibro replacement stone columns in soft clay. *In Geotechnics of Soft Soils—Focus on Ground Improvement. Glasgow, 3–5 September 2008.*

Greenwood, D.A. & Kirsch, K. 1984. Specialist Ground Treatment by Vibratory and Dynamic Methods. *Piling and Ground Treatment.* The institution of Civil Engineers: Tomas Telford, London.

Guetif, Z., Bouassida, M. & Debats, J.M. 2007. Improved soft clay characteristics due to stone column installation. *Computers and Geotechnics 34*(2): 104–111.

Kirsch, F. 2006. Vibro stone column installation and its effect on ground improvement. *In Proceedings of Numerical Modelling of Construction Processes in Geotechnical Engineering for Urban Environment, Bochum, Germany, 23–24 March 2006.* Taylor and Francis, London: 115–124.

Kirsch, K., Kirsch, F. 2010. *Ground improvement by deep vibratory methods.* New York: Spon Press.

Massarsch, K.R. 1991. Deep Soil compaction Using Vibratory Probes in Deep Foundation Improvement. STP1089 ASTM.

Priebe, H.J. 1995. Design of vibro replacement. *Ground Engineering 28*(10): 31–37.

Sonderman, W. & Wehr, W. 2004. Deep vibro techniques. *Ground Improvement.* 2nd Edition Ed. Moseley & Kirsch.

Installation Effects in Geotechnical Engineering – Hicks et al. (eds)
© 2013 Taylor & Francis Group, London, ISBN 978-1-138-00041-4

Numerical analyses of stone column installation in Bothkennar clay

J. Castro
University of Cantabria, Santander, Spain

M. Karstunen
Chalmers University of Technology, Gothenburg, Sweden
University of Strathclyde, Glasgow, UK

N. Sivasithamparam
University of Strathclyde, Glasgow, UK
Plaxis BV, Delft, The Netherlands

C. Sagaseta
University of Cantabria, Santander, Spain

ABSTRACT

The paper presents the results of numerical simulations studying the installation effects of stone columns in a natural soft clay. Stone column installation is modelled as an undrained expansion of a cylindrical cavity, using the finite element code PLAXIS that allows for large displacements. The properties of the soft clay correspond to Bothkennar clay, a soft Carse clay from Scotland (UK). The complexity of this material is simulated via two advanced recently developed constitutive formulations able to account for the soil structure, namely S-CLAY1 and S-CLAY1S. Modified Cam Clay model is also used for comparison purposes. The paper shows the new stress field and state parameters after column installation and the subsequent consolidation process. This sets the basis for including installation effects in studying the settlement reduction caused by stone columns.

REFERENCES

Arnold, M. & Herle, I. 2009. Comparison of vibro-compaction methods by numerical simulations. *Int. J. Num. Analytical Methods in Geomechanics* 33(16): 1823–1838.

Balaam, N.P. & Booker, J.R. 1985. Effect of stone column yield on settlement of rigid foundations in stabilized clay. *Int. J. Num. Analytical Methods in Geomechanics* 9(4): 331–351.

Barksdale, R.D. & Bachus, R.C. 1983. *Design and construction of stone columns*. National Technical Information Service Report FHWA/RD-83/026, Springfield, Virginia.

Brinkgreve, R.B.J. 2008. *Plaxis 2D—Version 9*. Plaxis, Delft.

Carter, J.P., Randolph, M.F. & Wroth, C.P. 1979. Stress and pore pressure changes in clay during and after the expansion of a cylindrical cavity. *Int. J. Num. Analytical Methods in Geomechanics* 3(4): 305–322.

Castro, J. & Karstunen, M. 2010. Numerical simulations of stone column installation. *Canadian Geotechnical Journal* 47(10): 1127–1138.

Castro, J. & Sagaseta, C. 2009. Consolidation around stone columns. Influence of column deformation. *International Journal for Numerical and Analytical Methods in Geomechanics* 33(7): 851–877.

Castro, J. & Sagaseta, C. 2012. Pore pressure during stone column installation. *Proc. of ICE—Ground improvement* 165(2): 97–109.

Egan, D., Scott, W. & McCabe, B. 2008. Installation effects of vibro replacement stone columns in soft clay. In Karstunen and Leoni (eds.) *Geotechnics of Soft Soils—Focus on Ground Improvement, Glasgow, 3–5 September 2008*. Taylor and Francis: London, pp. 23–29.

Elshazly, H.A., Hafez, D. & Mosaad, M. 2006. Back calculating vibro-installation stresses in stone columns reinforced grounds. *Proc. of ICE—Ground improvement* 10(2): 47–53.

Elshazly, H.A., Elkasabgy, M. & Elleboudy, A. 2008. Effect of inter-column spacing on soil stresses due to vibro-installed stone columns: interesting findings. *Journal of Geotechnical and Geological Engineering* 26: 225–236.

Gäb, M., Schweiger, H.F., Thurner, R. & Adam, D. 2007. Field trial to investigate the performance of a floating stone column foundation. In *Proc. of the 14th European Conf. Soil Mech. Geotech. Eng.*, Madrid, 24–27 September 2007. Millpress: Amsterdam, pp. 1311–1316.

Géotechnique Symposium in print. 1992. Bothkennar soft clay test site: characterisation and lessons learned. *Géotechnique* 42(2).

Guetif, Z., Bouassida, M. & Debats, J.M. 2007. Improved soft clay characteristics due to stone column installation. *Computers and Geotechnics* 34(2): 104–111.

Karstunen, M., Krenn, H., Wheeler, S.J., Koskinen, M. & Zentar, R. 2005. Effect of anisotropy and destructuration on the behaviour of Murro test embankment. *ASCE International Journal of Geomechanics* 5(2): 87–97.

Kirsch, F. 2004. *Experimentelle und numerische Untersuchungen zum Tragverthalten von Rüttelstopfsäulengruppen.* Dissertation, Technische Universität Braunschweig.

Kirsch, F. 2006. Vibro stone column installation and its effect on ground improvement. In *Proc. of Numerical Modelling of Construction Processes in Geotechnical Engineering for Urban Environment, Bochum, Germany, 23–24 March 2006.* Taylor and Francis: London, pp. 115–124.

Lee, F.H., Juneja, A. & Tan, T.S. 2004. Stress and pore pressure changes due to sand compaction pile installation in soft clay. *Géotechnique* 54(1): 1–16.

Massarsch, K.R. & Fellenius, B.H. 2002. Vibratory compaction of coarse-grained soils. *Canadian Geotechnical Journal* 39(3): 695–709.

McMeeking, R.M. & Rice, J.R. 1975. Finite-element formulations for problems of large elastic-plastic deformation. *International Journal of Solids and Structures* 11: 606–616.

Priebe, H.J. 1995. Design of vibro replacement. *Ground Engineering* 28(10): 31–37.

Installation Effects in Geotechnical Engineering – Hicks et al. (eds)
© *2013 Taylor & Francis Group, London, ISBN 978-1-138-00041-4*

Execution of Springsol® deep mixed columns: Field trials

S. Melentijevic, F. Martin & L. Prieto
Grupo Rodio-Kronsa S.L.U., Madrid, Spain

ABSTRACT

The reparation and underpinning of existing structures and infrastructure, due to different post-constructive pathologies, often needs the ground improvement of man-made fills and soft soils in general. The increasing growth of deep soil mixing methods has evolved into the development of the special Springsol® tool, which permits the installation of deep mixed soil-cement columns under existing superstructures by the application of new procedures linked with a controlled opening system. Based on the results of different recently executed field trial tests and projects, this paper describes some advantages of the Springsol® tool used for the construction of deep mixed columns. The execution parameters and methods of quality control during and after column installation are described and analyzed, taking into account installation effects and its influence on the geo-mechanical characteristics of the improved soil. Full scale test results are presented and compared with results obtained by laboratory tests on core samples.

REFERENCES

Bruce, D.A. 2001. An introduction to deep mixing methods as used in geotechnical applications, Volume III: The verification and properties of treated ground.
U.S. Department of Transportation, Federal Highway Administration, report FHWA RD-99–167.

CDIT (Coastal Development Institute of Technology), Japan. 2002. The Deep Mixing Method, A.A. Balkema.

Le Kouby, A., Bourgeois, E. & Rocher-Lacoste, F. 2010. Subgrade Improvement Method for Existing Railway Lines—an Experimental and Numerical Study. *EJGE* Vol. 15: 461–494.

Melentijevic, S., Prieto, L. & Arcos, J.L. 2012. Aplicaciones de columnas suelo-cemento tipo Springsol®. 9° Simposio Nacional de Ingeniería Geotécnica. Cimentaciones y Excavaciones Profundas. Proc. Symp., Sevilla, 17–19 October 2012: 175–189.

Rodriguez Abad, R. & Estaire Gepp, J. 2012. Determinación mediante WD-XRF del contenido de cemento en suelos inyectados y en mezclas de suelo-cemento. *9° Simposio Nacional de Ingeniería Geotécnica. Cimentaciones y Excavaciones Profundas. Proc. Symp., Sevilla, 17–19 October 2012:* 255–268.

UNE-EN15309:2007. Characterization of waste and soil. Determination of elemental composition by X-ray fluorescence.

Laboratorio de Geotecnia. Cedex. 2011. Informe de laboratorio para Grupo Rodio-Kronsa. Columnas suelo-cemento para el proyecto de investigación de nuevas técnicas y herramientas de soil-mixing.

Installation Effects in Geotechnical Engineering – Hicks et al. (eds)
© *2013 Taylor & Francis Group, London, ISBN 978-1-138-00041-4*

A method of modelling stone column installation for use in conjunction with unit cell analyses

B.G. Sexton & B.A. McCabe
College of Engineering and Informatics, National University of Ireland, Galway, Ireland

ABSTRACT

Traditionally, the majority of numerical studies investigating stone column behaviour have studied the problem under unit cell (axisymmetric) conditions, in which the granular columns tend to be 'wished-in-place' (no installation effects). In this study, cylindrical cavity expansion is used to work out post-installation lateral earth pressure coefficients (and hence a post-installation stress-regime in the ground) arising due to the lateral expansion and subsequent soil remoulding caused by the vibrating poker as columns are installed in a soft clay. Two sets of two-dimensional axisymmetric finite element analyses have then been carried out using PLAXIS 2D to examine load-settlement behaviour, the first set assuming the coefficient of earth pressure to be unaffected by column installation, while the second set have been conducted using increased earth pressure coefficients based on the cavity expansion procedure. The Hardening Soil Model (no viscous effects) has been used to model the behaviour of the granular column material and the soft clay. Settlement improvement factors calculated using both approaches have been compared to establish the effect of column installation. Predicted improvement factors have been put into context by comparison with existing analytical settlement design approaches. The results indicate that this approach can be used as a realistic means of accounting for column installation in conjunction with unit cell analyses, with larger improvement factors predicted when installation (increased K) is taken into account.

REFERENCES

Allman, M.A. & Atkinson, J.H. 1992. Mechanical properties of reconstituted Bothkennar soil. *Géotechnique* 42(2): 289–301.

Ambily, A.P. & Gandhi, S.R. 2007. Behavior of Stone Columns Based on Experimental and FEM Analysis. *Journal of Geotechnical and Geoenvironmental Engineering* 133(4): 405–415.

Balaam, N.P. & Booker, J.R. 1981. Analysis of rigid rafts supported by granular piles. *International Journal for Numerical and Analytical Methods in Geomechanics* 5(4): 379–403.

Brinkgreve, R.B.J., Swolfs, W.M. & Engin, E. 2011. *PLAXIS 2D 2010 Material Models Manual*, PLAXIS B.V.

Carter, J.P., Randolph, M.F. & Wroth, C.P. 1979. Stress and pore pressure changes in clay during and after the expansion of a cylindrical cavity. *International Journal for Numerical and Analytical Methods in Geomechanics* 3(4): 305–322.

Castro, J. & Karstunen, M. 2010. Numerical simulations of stone column installation. *Canadian Geotechnical Journal* 47(10): 1127–1138.

Castro, J. & Sagaseta, C. 2009. Consolidation around stone columns. Influence of column deformation. *International Journal for Numerical and Analytical Methods in Geomechanics* 33(7): 851–877.

Debats, J.M., Guetif, Z. & Bouassida, M. 2003. Soft soil improvement due to vibro-compacted columns installation. *Proceedings of the International Workshop "Geotechnics of Soft Soils. Theory and Practice"*, Noordwijkerhout, 551–556.

Domingues, T.S., Borges, J.L. & Cardoso A.S. 2007. Stone columns in embankments on soft soils. Analysis of the effects of the gravel deformability. *Proceedings of the 14th European Conference on Soil Mechanics and Geotechnical Engineering*, Madrid, 1445–1450.

Elshazly, H., Hafez, D. & Mossaad, M. 2008. Back-calculating vibro-installation stresses in stone-column-reinforced soils. *Proceedings of the ICE—Ground Improvement* 10(2): 47–53.

Gäb, M., Schweiger, H.F., Kamrat-Pietraszewska, D. & Karstunen, M. 2008. Numerical analysis of a floating stone column foundation using different constitutive models. *Proceedings of the 2nd International Workshop on the Geotechnics of Soft Soils—Focus on Ground Improvement*, Glasgow, 137–142.

Gibson, R.E. & Anderson, W.F. 1961. In situ measurements of soil properties with the pressure meter. *Civil Engineering and Public Works Review* 56 (658): 615–618.

Goughnour, R.R. & Bayuk, A.A. 1979. A Field Study of Long Term Settlements of Loads Supported by Stone Columns in Soft Ground. *Proceedings of the International Conference on Soil Reinforcement: Reinforced Earth and Other Techniques (Coll. Int. Renforcements des Sols.)*, Paris, 1: 279–285.

Guetif, Z., Bouassida, M. & Debats, J.M. 2007. Improved soft clay characteristics due to stone column installation. *Computers and Geotechnics* 34(2): 104–111.

Killeen, M.M. & McCabe, B.A. 2010. A Numerical Study of Factors Affecting the Performance of Stone Columns Supporting Rigid Footings on Soft Clay. *Proceedings of the 7th European Conference on Numerical Methods in Geotechnical Engineering*, Trondheim, 833–838.

Kirsch, F. 2006. Vibro Stone Column Installation and its Effect on Ground Improvement. *Proceedings of the International Conference on Numerical Modelling of Construction Processes in Geotechnical Engineering for Urban Environment*, Bochum, 115–124.

McCabe, B.A., Nimmons, G.J. & Egan, D. 2009. A review of field performance of stone columns in soft soils. *Proceedings of the ICE—Geotechnical Engineering* 162(6): 323–334.

Nash, D.F.T., Powell, J.J.M. & Lloyd, I.M. 1992. Initial investigations of the soft clay test site at Bothkennar. *Géotechnique* 42(2): 163–181.

Priebe, H.J. 1976. Evaluation of the settlement reduction of a foundation improved by Vibro-Replacement. *Bautechnik* 2: 160–162.

Priebe, H.J. 1995. The design of vibro replacement. *Ground Engineering* 28(10): 31–37.

Pulko, B., Majes, B. & Logar, J. 2011. Geosynthetic-encased stone columns: Analytical calculation model. *Geotextiles and Geomembranes* 29(1): 29–39.

Schanz, T., Vermeer, P.A. & Bonnier, P.G. 1999. The hardening soil model: Formulation and verification. *Beyond 2000 in Computational Geotechnics—Ten Years of PLAXIS International*, Amsterdam, 281–290.

Vesic, A.S. 1972. Expansion of cavities in infinite soil masses. *Journal of the Soil Mechanics and Foundations Division* 98(4): 265–290.

Watts, K.S., Johnson, D., Wood, L.A. & Saadi, A. 2000. An instrumented trial of vibro ground treatment supporting strip foundations in a variable fill. *Géotechnique* 50(6): 699–708.

Installation Effects in Geotechnical Engineering – Hicks et al. (eds)
© *2013 Taylor & Francis Group, London, ISBN 978-1-138-00041-4*

Cement grout filtration in non-cohesive soils

X.A.L. Stodieck
Federal Waterways Engineering and Research Institute, Karlsruhe, Germany
Norwegian University of Science and Technology, Trondheim, Norway

T. Benz
Norwegian University of Science and Technology, Trondheim, Norway

ABSTRACT

This paper is concerned with cement grout filtration during installation of pressure grouted ground anchors in non-cohesive soils. It is aimed for a simple model that can predict the build up of filter cake during grouting. A series of laboratory tests were carried out to investigate the time dependent expulsion of water from cement grouts at different grouting pressures and water contents. Additionally, permeability tests were performed on filter cake material. Different approaches to model the filtration process of cement grouts are compared and applied in back calculation of the performed tests. Material parameters such as permeability- and consolidation coefficients are provided for the cement grouts tested. A method to estimate the time required for filter cake formation is proposed.

REFERENCES

Abuel-Naga, H.M. & Pender, M.J. (2012). Modified terzaghi consolidation curves with effective stress-dependent coefficient of consolidation. *Gotechnique Letters* 2(2), 43–48.

Carman, P.C. (1956). *Flow of gases through porous media.* London: Butterworths Scientific Publications.

Casagrande, A. & Fadum, R.E. (1940). *Notes on soil testing for engineering purposes.* Cambridge, Mass.: Harvard University, Graduate School of Engineering.

Kleyner, I. & Krizek, R.J. (1995).MathematicalModel for Bore-Injected Cement Grout Installations. Journal of geotechnicalengineering. 121(11), 782–788.

Lee, S.-W., T.-S. Kim, B.-K. Sim, J.-S. Kim, & I.-M. Lee (2012). Effect of pressurized grouting on pullout resistance and group efficiency of compression ground anchor. *Can. Geotech. J. Canadian Geotechnical Journal 49*(8), 939–953.

McKinley, J.D. (1993). *Grouted ground anchors and the soil mechanics aspects of cement grouting.* Ph.D. thesis.

Picandet, V., Rangeard, D., Perrot, A. & Lecompte, T. (2011). Permeability measurement of fresh cement paste. *Cement and Concrete Research 41*(3), 330–338.

Taylor, D.W. (1948). *Fundamentals of soil mechanics.* New York: J. Wiley.

Warner, J. (2004). *Practical Handbook of Grouting—Soil, Rock, and Structures.* Norwich, NY: JohnWiley & Sons.

The undrained cohesion of the soil as a criterion for column installation with a depth vibrator

J. Wehr

Keller Holding GmbH, Offenbach, Germany

ABSTRACT

Piles, vibro concrete columns and rigid inclusions cannot be installed in liquid media like water with $c_u = 0$ kPa, because a cone shaped slope will form instead of a cylindrical body. In international standards a limit of $c_u = 15$ kPa is currently used as a lower boundary. Additionally the minimum center to center distance between fresh concrete displacement piles without permanent casing is specified in EN 12699 between 6–10 times the diameter depending on the c_u-value. These limits should be applied to all kinds of displacement piles like rigid inclusions in order to avoid damaging neighbouring columns during installation.

However, recent world-wide site experiences reveal that these limits are not valid for granular columns like vibro stone columns. The limit of the undrained cohesion should be reduced to $c_u = 5$ kPa.

REFERENCES

Borchert, K.-M. & Kattner, G. 2004. *Prüfbericht der Probebelastungen an 3 Fertigmörtelstopfsäulen*, internal report. FGFS, Forschungsgruppe für Straßenwesen. 1979. *Merkblatt für die Untergrundverbesserung durch Tiefenrüttler.*

Gundacker, S. 2004. *Anwendungsgrenzen bei der Herstellung von Schottersäulen*, Diploma-thesis, Technical university of Kaiserslautern.

Marte, R. & Schuller, H. 2005. Verbesserung sehr weicher Seesedimente und Torfe durch Schottersäulen—zwei Fallbeispiele, *Bauingenieur*, Volume 80, 430–440.

Perlea, V. 2000. Liquefaction of cohesive soils, soil dynamics and liquefaction, *ASCE geotechnical special publication*, no. 107, 58–76.

Raju, V.R. 1997. *The behaviour of very soft cohesive soils improved by vibro replacement*, ground improvement conference, London.

Raju, V.R. & Hoffmann, G. 1996. *Treatment of tin mine tailings in Kuala Lumpur using vibro replacement*. Proc. 12th SEAGC, May 1996, Kuala Lumpur.

Raju, V.R. 2002. *Vibro replacement for high earth embankments and bridge abutment slopes in Putrajaya, Malaysia*, International Conference on Ground Improvement Techniques, Malaysia, p. 607–614.

Völzke, B. 2001. *Böschungssicherung für die Sanierung des Metallhüttengeländes in Lübeck-Herrenwyk*, 8. Darmstädter Geotechnik Kolloquium, 233–239.

Zimmermann, K.-U. 2003. Gründung von Verkehrswegebauten in Feuchtgebieten mit organischen Böden geringer Scherfestigkeit, *Mitteilungen des Instituts für Grundbau und Bodenmechanik, Universität Braunschweig*, Pfahlsymposium, issue 71, 71–81dense sand, Can. Geotech. J. 33, 209–218.

Soil-structure interaction

Modeling of rock fall impact using Discrete Element Method (DEM)

G. Grimstad & O. Melhus
Oslo and Akershus University College of Applied Sciences, Norway

S. Degago & R. Ebeltoft
The Norwegian Public Roads Administration, Norway

ABSTRACT

Buried corrugated steel culverts, cushioned on top by a layer of granular material, are sometimes used to protect infrastructure from rock falls. A crucial point in design of such structure is the cushion material ability to absorb the kinetic energy from the rock block. This is evaluated by two important parameters, i.e. the resulting impact load and penetration of the falling rock. In this article a Discrete Element Model (DEM) is used in simulation of the impact from a falling rock block on such a cushion layer. The simulation is compared to measurement data from a full-scale field test. Applicability and potential of DEM to simulate rock fall impacts on cushion are assessed and discussed in the light of measurements from full scale test and estimations using empirical relations proposed by Norwegian guideline.

REFERENCES

ASTRA, 2008. Einwirkungen infolge Steinschlags auf Schutzgalerien. *Richtlinie, ASTRA 12 006*, V2.03.

Cundall, P.A. & Strack, O.D.L. 1979. A discrete numerical model for granular assemblies. *Géotechnique*, 29: 47–65.

Degago, Samson, A. 2007. Impact tests on sand with numerical modeling, emphasizing on the shape of the falling object, *Master thesis*, NTNU, Trondheim, Norway.

Degago, S.A, Ebeltoft, R. & Nordal, S. 2008. Effect of rock fall geometries impacting soil cushion: A numerical procedure. *The 12th International Conference of International Association for Computer Methods and Advances in Geomechanics (IACMAG)*, Goa, India.

Ebeltoft, R. & Larsen, J.O. 2006. Instrumentation of buried flexible structure subjected to rock fall loading. *Joint international conference on computing and decision making in civil and building engineering.* Montreal, Canada.

Ebeltoft, R., Gloppestad, J. & Nordal, S. 2006. Finite element analysis of buried flexible culverts subjected to rockfall loading. *Numerical Methods in Geotechnical Engineering, NUMGE06*, Graz, Austria.

Gupta, A. & Saigal K.R. 2003. Simple Formulation to Evaluate Surface Impacts on Buried Steel Pipelines, *Vessel and Piping Stresses, Welding research Council Bulletin no. 479.*

Imre, B., Räbsamen, S. & Springman, S.M. 2008. A coefficient of restitution of rock materials. *Computers & Geosciences,* 34(4): 339–350.

Jacquemound, J. 1999. Swiss Guideline for the Design of Rockfall Protection Galleries: Background, Saftey concept and case histories, *Joint Japan-Swiss Scientific Seminar on Impact Load by Rock Falls and Design of Protection Structures*, Kanazawa, Japan, 95–102.

Kawahara, S. & Muro, T. 2006, Effects of Dry Density and Thickness of Sandy soil on Impact Response due to Rock-fall, *Journal of Terramechanics*, 43: 329–340.

Kishi, N., Ikeda, K., Mikami, H. & Takemoto S. 1999. A proposed New Design Procedure for RC Rock-Sheds, *Joint Japan-Swiss Scientific Seminar on Impact Load by Rock Falls and Design of Protection Structures*, 103–112.

Labiouse, V., Descoueudres, E. & Montani S. 1995, Numerical Analysis of Rock Blocks Impact a soil Cushion, *Numerical Models in Geomechnics-NUMOG V*, Pande & Pi-eetruszczak (eds), 645–650.

Masuya, H., Tanak, Y., Onda, S. & Ihara, T. 1999, Evaluation of Rock Falls on Slopes and Simulations of the Motion of Rock Falls in Japan, *Joint Japan-Swiss Scientific Seminar on Impact Load by Rock Falls and Design of Protection Structures*, Kanazawa, Japan, 21–28.

Montani, Stoffel, S., Labouse, V. & Descoueudres, F. 1999, Action of Falling Blocks Impacting Rocksheds Covered with a Soil Cushion, *Joint Japan-Swiss Scientific Seminar on Impact Load by Rock Falls and Design of Protection Structures*, Kanazawa, Japan, 51–57.

Pichler, B., Hellmich, C. & Mang, H.A. 2005. Impact of rocks onto gravel, design and evaluation of experiments. *International Journal of Impact Engineering* 31(5): 559–578.

Plassiard, J.-P. & Donz´e, F.-V. 2009. Rockfall Impact Parameters on Embankments: A Discrete Element Method Analysis, *Struct. Eng. Int.*, 19: 333–341.

Rojek, J. & Oñate, E., 2008. Multiscale analysis using a coupled discrete/finite element model. *Interaction and Multiscale Mechanics*, 1(1): 1–31.

SVV, 2011. Sikring av veger mot steinskred, Grunnlag for veiledning, *VD rapport*, No. 32.

Timoshenko, S. 1951. *Theory of Elasticity*. McGraw-Hill Book Company.

Volkwein, A., Schellenberg, K., Labiouse, V., Agliardi, F., Berger, F., Bourrier, F., Dorren, L.K.A., Gerber, W. & Jaboyedoff, M. Rockfall characterisation and structural protection—a review. *Nat. Hazards Earth Syst. Sci.*, 11: 2617–2651.

Yoshida, H., Masuya, H. & Ihara, T. 1988, Experimental Study of Impulsive Design Load for Rock Sheds, *IABSE Proceedings P-127/88*, 61–74.

Installation Effects in Geotechnical Engineering – Hicks et al. (eds)
© *2013 Taylor & Francis Group, London, ISBN 978-1-138-00041-4*

Investigation into the factors affecting the shaft resistance of driven piles in sands

D. Igoe, K. Gavin & L. Kirwan

School of Civil, Structural and Environmental Engineering, University College Dublin, Dublin, Ireland

ABSTRACT

The paper presents the results of field tests performed to study the effects of the installation technique, degree of plugging, cyclic loading and ageing on the shaft resistance developed on open-ended piles in sand. Two instrumented model piles were jacked and driven into an artificially created loose sand deposit in Blessington, Ireland. The results from these tests indicated that the equalized radial effective stresses which are suggested to control the shaft capacity vary strongly with the degree of plugging and number of load cycles experienced during installation. A comparison of jacked and driven installations suggest similar radial stresses were developed provided the jacked pile had experienced a sufficient number load cycles. The degree of plugging experienced during installation controlled the radial stresses near the bottom of the pile, with closed-ended or plugged piles developing high stresses near the pile base and exhibiting friction fatigue up the shaft, compared with open-ended coring piles which exhibited relatively low stresses along the length of the pile shaft. A comparison with full scale 340 mm diameter pipe piles driven into the dense sand in Blessington noted comparable radial stresses when the pile was fully coring but exhibited a larger increase in radial stress near the pile toe as the pile became plugged. Further tests on the effects of ageing show a pile shaft capacity increase of 260% over 220 days after driving. Further research is underway in Blessington to investigate the mechanisms controlling this ageing behaviour.

REFERENCES

Chow, F., Jardine, R.J., Brucy, F. and Nauroy, J.F. 1997. "Time related increases in the shaft capacities of driven piles in sand," *Geotechnique*, Vol. 47, No. 2, pp. 353–361.

Gavin, K.G. and Lehane, B.M., 2003. "The shaft capacity of pipe piles in sand," *Canadian Geotechnical Journal*, Vol. 40, No., pp. 36–45.

Gavin, K.G., Igoe, D. and Doherty, P., 2011. "Use of open-ended piles to support offshore wind turbines: A state of the art review," *Proceedings of the ICE—Geotechnical Engineering*, Vol. 164, No. GE4, pp. 245–256.

Gavin, K.G, Igoe, D. and Kirwan, L., 2013. "The effect of ageing on the axial capacity of piles in sand", *ICE—Geotechnical Engineering Special Edition*, In Press.

Heerema, E., 1980. "Predicting pile driveability: heather as an illustration of the friction fatigue theory," *Ground Engineering*, Vol. 13, No., pp. 15–37.

Igoe, D., Doherty, P. and Gavin, K.G., 2010. "The development and testing of an instrumented open-ended model pile," *Geotechnical Testing Journal*, Vol. 33, No. 1, pp. 1–11.

Igoe, D., Gavin, K.G. and O'Kelly, B.C., 2011. "The shaft capacity of open-ended piles in sand," *Journal of Geotechnical and Geoenvironmental Engineering, ASCE*, Vol. 137, No. 10, pp. 903–913.

Jardine, R.J., Chow, F.C., Overy, R.F. and Standing, J., 2005. ICP Design Methods for Driven Piles in Sands and Clays. T. Telford. London, University of London (Imperial College).

Jardine, R.J. and Chow, F.C., 2007. "Some Recent Developments in Offshore Pile Design," *6th International Offshore Site Investigation and Geotechnics Conference*, London.

Lehane, B.M. 1992. "Experimental investigations of pile behaviour using instrumented field piles," PhD Thesis, University of London (Imperial College).

Lehane, B.M. and Jardine, R.J., 1994. "Shaft capacity of driven piles in sand: a new design approach," *Conference on the Behaviour of Offshore Structures*, Boston, Mass.

Lehane, B.M., Schneider, J.A. and Xu, X., 2005. "The UWA-05 method for prediction of axial capacity of driven piles in sand," *Frontiers in Offshore Geotechnics: ISFOG*, Perth, University of Western Australia.

White, D.J., Schneider, J.A. and Lehane, B.M., 2005. "The influence of effective area ratio on shaft friction of displacement piles in sand," *Frontiers in Offshore Geotechnics, ISFOG*, University of Western Australia, Perth.

Installation Effects in Geotechnical Engineering – Hicks et al. (eds)
© *2013 Taylor & Francis Group, London, ISBN 978-1-138-00041-4*

Monitoring and risk assessment in EPB TBM's in urban environments: High speed railway tunnel Sants-Sagrera running next to Sagrada Familia Basilica (World Heritage)

J.E. París Fernández & J. Gómez Cabrera
Sener Ingeniería y Sistemas, S.A., Barcelona, Spain

ABSTRACT

In this paper we are going to summarize the procedures used to manage the operation of a TBM (Tunnel Boring Machine) type EPB (Earth Pressure Balance) with continuous and "real time" monitoring of working and parameters. To show the characteristics of this method we chose a relevant example, Sagrada Familia Basilica designed by Antoni Gaudí, which is a UNESCO World Heritage Site. Some control parameters have been followed in order to detect possible damage to the building structure. During this time, Sener Ingenieria y Sistemas supported the project with direct monitoring and risk assessment to the TBM, reaching an advanced level in order to detect anomalies in the TBM operation, ground movements or building failure.

REFERENCES

Anagnostou, G. & Kovari, K. 1994. Stability analysis for tunneling with slurry and EPB shields. *Mir 94 "Gallerie in condizioni difficilli"*, Torino.

Gómez, J. 2009. Excavation monitoring in tunnels execute by EPB. SENER's experience in Oporto and Lisboa metro lines. *Obras Urbanas. Febrero 2009, n° 13.*

Gómez, J. 2011. Follow up and risks control in TBM EPB operation in "real time". Sants-Sagrera Tunnel. *I Foro Internacional Ferroviario. Bcn Rail 2011*, Barcelona.

Gomez, J. & Roldan, J. 2012. Follow up and risk assessment in EPB TBM operationsin urban environments. Sants-Sagrera tunnel crossing underneath Sagrada Familia Temple. *Second Colombian and First Andean and Central American Congress and Exhibition of No–Dig Technologies and Underground Infrastructure 2012*, Cartagena de Indias, Colombia.

Effect of roughness on keying of plate anchors

D. Wang, C. Han & C. Gaudin

Centre for Offshore Foundation Systems, The University of Western Australia, Crawley, WA, Australia

ABSTRACT

Suction Embedded PLate Anchors (SEPLAs) are a relatively new type of anchorage in deep waters. The SEPLA is inserted vertically into clayey seabed and then pulled to rotate until it becomes nearly perpendicular to the loading inclination. The keying response has been studied by means of centrifuge tests and Large Deformation Finite element (LDFE) analyses. However, the predicted ultimate losses of embedment were lower than the experimental results in most scenarios. The effect of anchor roughness on the keying process is then investigated using LDFE approach with contact algorithm. A friction coefficient of 0.3 in typical normally consolidated kaolin clays is determined by comparing the numerical results with three groups of experimental data measured. The loss of embedment during keying depends on the combined influences of anchor roughness, anchor thickness and loading eccentricity. When the anchor thickness ratio is less than 0.7 and the loading eccentricity ratio not larger than 0.5, the anchor roughness needs to be considered.

REFERENCES

Cassidy, M.J., Gaudin, C., Randolph, M.F., Wong, P.C., Wang, D. & Tian, Y. 2012. A plasticity model to assess the keying of plate anchors. *Geotechnique*, 62, 825–836.

Chen, W. & Randolph, M.F. 2007. External radial stress changes and axial capacity for suction caissons in soft clay. *Geotechnique*, 57, 499–511.

Dingle, H.R.C., White, D.J. & Gaudin, C. 2008. Mechanisms of pipe embedment and lateral breakout on soft clay. *Canadian Geotechnical Journal*, 45, 636–652.

Gaudin, C., O'Loughlin, C.D., Randolph, M.F. & Lowmass, A.C. 2006. Influence of the installation process on the performance of suction embedded plate anchors. *Geotechnique*, 56, 381–391.

Gaudin, C., Simkin, M., White, D.J. & O'Loughlin, C.D. 2010. Experimental investigation into the influence of a keying flap on the keying behaviour of plate anchor. *Proc. 20th Int. Offshore and Polar Engineering Conf.*, 533–540.

Gaudin, C., Tham, K.H. & Ouahsine, S. 2008. Plate anchor failure mechanism during keying process. *Proc. 18th Int. Offshore and Polar Engineering Conf.*, 613–619.

O'Loughlin, C.D., Lowmass, A., Gaudin, C. & Randolph, M.F. 2006. Physical modelling to assess keying characerics of plate anchors. *Proce. of 6th Int. Conf. on Physical Modelling in Geotechnics*, 659–665.

Randolph, M.F., Wang, D., Zhou, H., Hossain, M.S. & Hu, Y. 2008. Large Deformation Finite Element Analysis for Offshore Applications. *Proc. 12th Int. Conf. International Association for Computer Methods and Advances in Geomechanics*, 3307–3318. Simulia. 2010. ABAQUS 6.10 Manuals. *Simulia*.

Song, Z., Hu, Y., O'Loughlin, C. & Randolph, M.F. 2009. Loss in Anchor Embedment during Plate Anchor Keying in Clay. *Journal of Geotechnical and Geoenvironmental Engineering*, 135, 1475–1485.

Wang, D., Gaudin, C. & Randolph, M.F. 2012. Large deformation finite element analysis investigating the performance of anchor keying flap. *Ocean Engineering*, under revision.

Wang, D., Hu, Y. & Randolph, M.F. 2010. Three-dimensional large deformation finite-element analysis of plate anchors in uniform clay. *Journal of Geotechnical and Geoenvironmental Engineering*, 136, 355–365.

Wang, D., Hu, Y. & Randolph, M.F. 2011. Keying of Rectangular Plate Anchors in Normally Consolidated Clays. *Journal of Geotechnical and Geoenvironmental Engineering*, 137, 1244–1253.

White, D.J., Take, W.A. & Bolton, M.D. 2003. Soil deformation measurement using particle image velocimetry (PIV) and photogrammetry. *Geotechnique*, 53, 619–631.

Wilde, B., Treu, H. & Fulton, T. 2001. Field testing of suction embedded plate anchors. *Proc. 11th Offshore and Polar Engineering Conf.*, 544–551.

Installation Effects in Geotechnical Engineering – Hicks et al. (eds)
© *2013 Taylor & Francis Group, London, ISBN 978-1-138-00041-4*

Author index